Preface

"Highland Animals" is the second volume to be published in our Highland Life series. The first, "Highland Birds" by Desmond Nethersole-Thompson, confirmed our view that a useful and authoritative Highland library would be welcomed by the wider community.

For this new volume we commissioned David Stephen, known throughout Britain as an authority on wild life and a writer of exceptional ability. I am sure his work will have a warm welcome both from visitors to the Highlands and from the local inhabitants. There are more animals in the Highlands than most people think, and we ought to know more about them. It is much easier to see an osprey than a badger nowadays; but for my money the badger is even more interesting. David Stephen will be your guide to badgers and to a hundred other wild creatures of the Highlands.

Andrew Gilchrist

Chairman · Highlands & Islands Development Board

Foreword

When the Highlands and Islands Development Board invited me, in 1972, to write this book I accepted at once—partly because of my long love-affair with the Highlands, partly because mammals are my main interest, and partly because I liked what the Board had done, and is doing, for the area.

It is generally agreed that mammals are more difficult to work with than birds, which is why photographs of Highland mammals are so much scarcer than those of birds.

Writing this book was, for many reasons, a great pleasure to me. One reason was the happy cooperation between Gordon Lyall of the Highlands and Islands Development Board, Bill MacKay, the imaginative designer and myself. They gave me unlimited scope, and I record my thanks to both.

A man doesn't write a book like this without the help of a number of people. I would like to acknowledge my indebtedness to my old friend and colleague, Dr. J. D. Lockie, with whom I have long worked in the field. My thanks are also due to Dr. Adam Watson and Ray Hewson of the Mountain and Moorland Ecology Unit at Banchory; to Dr. David Jenkins and Dr. J. Morton Boyd of The Nature Conservancy. For much field assistance over the years, I would like to thank Fergus Ferguson, Head Stalker to the Duke of Atholl. Lastly, there is my long-standing debt to my former English Master and very dear friend, the late J. Harrison Maxwell, who nurtured my love of the Highlands and its wildlife, and who pointed the way for me when I was a very small boy.

I consulted many books and papers during the writing of this book. I think it is a truism to say that no naturalist can get along today without Professor J. Ritchie's "The Influence of Man on Animal Life in Scotland". One of the very useful moderns is "Ecology and Land Use in Upland Scotland" by D. N. McVean and J. D. Lockie. But the modern classic is still, I think, "The Highlands and Islands" by Sir Frank Fraser Darling and Dr. J. Morton Boyd.

David Stephen

Introduction

Up to the Second World War, the Highlands and Islands were still pretty much of a closed book to people living outside; for two hundred years before that they were a book that successive Governments would have liked to see closed. Generations of politicians worked their lips to death in the Highlands' service, hiding behind mountains of paper while the real mountains were being degraded by misuse, and a culture and a way of life were being eroded away.

Hail sacred cow of economic growth! One day thou shalt find quiet pasture in the Celtic twilight.

That was the hope—I say hope because I am one of those unrepentants who still believe that the run-down of this unique area was originally a definite policy. But it wasn't quite achieved—despite depopulation, deforestation, the coming of the sheep, the creation of wet deserts, and intensive game preservation.

The coming of the sheep led to the going of the people and the brutality of The Clearances is a matter of history. The sheep in their turn were, to a great extent, dispossessed by deer and grouse when the new class of wealthy industrialists from the South became interested in sport. And when sport became a major form of land-use, the destruction of wildlife really began.

Of course there had been losses long before then. The giant Irish Elk, the pride of the Pleistocene, probably died out before it could meet a human being in Scotland. It was a magnificent animal, growing a spread of antlers weighing up to nineteen pounds; but it had to go. The lemming, the northern lynx and the rat vole probably disappeared at the dawn of the Christian era. The northern lynx, it is reckoned, held out until man came to push it over the edge into oblivion.

The great Elk, the deer of shambling gait, standing as tall as a man, and with a man's arms' spread of antlers, was gradually eliminated by the destruction of the forests and by hunting. It is difficult to realise that we still had it when Wallace marched to Stirling. There are a number of references to the great Elk, not to be confused with the Irish Elk or Giant Fallow Deer. In an old poem—"*Bas Dhiarmid*", we read:

Glen Shee, that glen by my side
Where oft is heard the voice of deer and elk

The brown bear disappeared earlier, probably in the ninth or tenth century, at which time serious destruction of the forest had hardly begun. The Romans knew the bear, and it is certain that Caledonian bears appeared in their brutal circuses at Rome. Martial wrote:

Nuda Caledonio sic pectora praebuit urso,
Non falsa pendens in cruce, Laureolus.

(Hanging on no slim cross, Laureolus
His naked body to a Caledonian bear
Thus proferred.)

The bear was the great Magh-Ghamhainn of the Gael—the "paw calf" of Highland legend, and still lives in place names like *Ruigh-na-beiste* (Monster's Brae) and *Toll-nam-beiste* (Hole of the Monsters). Professor Ritchie has stated that man must have exterminated the bear, because there were no changes in either climate or food supply to account otherwise for its disappearance.

The beaver, hunted for its skin, was probably exterminated by man. It seems likely that it managed to eke out an existence in the Highlands, probably in Inverness-shire, until the fifteenth or sixteenth century.

The wild boar was still alive in the sixteenth century and was greatly prized as a sporting animal. It was hunted by royalty and nobility. As a truly wild animal, it seems to have disappeared from Scotland about the same time as James VI. Attempts were made to reintroduce it, but failed. Besides these we also lost, in historic times, the wild ox, the native pony, the reindeer and the wolf.

The wolf outlasted them all, right into the eighteenth century. He was still howling when Montrose made his winter foray over the Corrieyarick to maul Argyll at Inverlochy. Some of the men who followed the Stuart standard to Sheriffmuir in 1715 would know him as a wild animal; certainly many of their fathers would. Yet it is unlikely that any of these men had ever seen a rabbit or a blackface sheep. For that matter, it is unlikely that any Scottish wolf had ever seen a rabbit.

When and where was the last Scottish wolf killed?

The story that ought to be true, and may well be, is that told of Macqueen of *Pall-a-Chrocain,* on the Findhorn, stalker to the Mackintosh of Mackintosh. Macqueen is usually given the credit for slaying the last wolf in Caledonia which, at that time, was not so stern or wild as man has since made it. The year was 1743—a bare two years before the last of the Stuarts raised his standard at Glenfinnan.

When the black beast, as the wolf was called, had killed two children, the Laird of Mackintosh called a *tainchel* to round him up, and Macqueen was the leader of the expedition sent against him.

Macqueen's story is surely the shortest and most remarkable animal story in the literature. Said he:

"As I came through the slochk by east the hill there, I foregathered wi' the beast. My long dog there turned him. I buckled wi' him, and dirkit him, and syne whuttled his craig, and brought awa his countenance for fear he should come alive again, for they are precarious creatures . . ."

It will be noted that the loss of all these big mammals took place before Scotland ceased to be a nation except, it seems, for the wolf who managed to hang on long enough to become British.

The next killing times arrived in the nineteenth century—the century of the game preservers' ascendancy.

Gamekeepers with traps, snares, nets, guns and poison waged war against every species considered inimical to game. Mammals and birds suffered. The pine marten was almost exterminated; the polecat disappeared (although there is a question-mark against this one). The wildcat was thinned down. Here are the often quoted figures for kills of predators in Glengarry for the years 1837 to 1840.

Foxes	11
Wildcats	198
House cats	78
Pine martens	246
Polecats	106
Weasels and Stoats	301
Badgers	67

The remarkable fact about these figures, however suspect they may be, is the proportion of foxes to wildcats, martens and polecats. The fox, it seems, was not so common in those days. There is no part of the Highlands where kills in this proportion could be achieved today. The fox would win out of the field. Glengarry is not an isolated example; it was not a unique nineteenth-century den of iniquity; it was merely down to the standard of its time.

Have the standards improved? The answer must be a qualified Yes. Most of the concessions have been made to the birds, and the Protection of Birds Act 1954 is the canon on this subject. The Royal Society for the Protection of Birds is active in Scotland and has done great service for its bird life.

The mammals have not been so lucky. Red deer and roe deer now have Close Seasons in law. The grey seal is protected, except for an annual cull permitted in Orkney. From 1st January 1974 the badger has been given legal protection of a sort.

But the story of the carnivores looks very like a continuation of the old, old one. Foxes, of course, are number one on the list, and are killed at all times by any means. Otters are protected in some parts, tolerated in others, and destroyed in others. Wildcats are generally killed on sight; or by shooting or trapping. Stoats and weasels are killed as a routine. The pine marten is not usually killed deliberately, but is sometimes caught in traps set for foxes. The gin trap, the one they get caught in, became illegal on 1st April, 1973. Polecats that may, or may not be, genuine polecats, fall to traps, or are shot at sight. Many keepers still kill hedgehogs as a routine.

In the context of wildlife conservation, as in so many others, the creation of The Nature Conservancy was the biggest single event of this, or any other, century. Here at last was a body with the power, the finance, and the know-how to carry out basic research and to plan management policies. Part of their remit was to establish National Nature Reserves, and we are now in the happy position where more than three-quarters of the land devoted to Nature Reserves in Britain is in Scotland, and the greater part of this is in the Highlands and Islands.

The Ben Eighe Reserve totals 10,000 acres, and is a traditional stronghold of the pine marten. The Cairngorms Reserve, including Glenfleshie, totals 58,800 acres. North Rona, is the main breeding station of the grey seal. The National Trust for Scotland owns St. Kilda which is also a National Nature Reserve. Rhum is a National Nature Reserve and Pat Lowe and Brian Mitchell have worked on red deer there. The island is now a major out-door Research Station. Rhum is of special interest because it was there that Fraser Darling wanted to do his red deer research more than thirty years ago. He was not allowed to do so. The first thing The Nature Conservancy did was to institute that very research.

Without research, there can be no planning; only guess work. Many people are now working on mammals in Scotland. Ted Smith did the original work on the Grey Seal, which was carried on by Morton Boyd of the Nature Conservancy, and James Lockie, now of Edinburgh University. Lockie and I have worked on foxes in the Highlands. Ray Hewson from Banchory is doing so today. The Mountain and Moorland Unit at Banchory tries to cover the whole spectrum of land-use and wildlife in their area. Adam Watson has been researching grouse and mountain hares. David Jenkins, of The Nature Conservancy, has been gathering information on the distribution and status of the wildcat. James Lockie and Dick Balharry have worked on the pine marten. Lockie and I have worked on stoats and weasels; in fact, the mustelids are one of my main interests. There is plenty of work going on. A great deal more requires to be done, and places like Rhum and Banchory need to be multiplied.

More recently the Forestry Commission has shown itself in a happier light, especially in its treatment of roe deer and predators. It has gone out of its way to make its forests safe for badgers; but it still permits the snaring of foxes and these snares often catch badgers. They also catch roe deer by the feet. Unfortunately, the Forestry Commission is still dedicated to monoculture—in this case the sitka spruce. Fraser Darling has called monoculture "a whirlpool, involving epidemic outbreaks of pests, followed by aerial spraying of lethal chemicals that kill fish," upset the calcium metabolism of raptorial birds and pollute the ecosystem.

The Highlands and Islands are still a great reservoir of wildlife—one of the greatest in the world. The Highlands and Islands Development Board, the body responsible for books such as this, is concerned very largely with economic growth and conservation; but it is also very much concerned with the wildlife resources that are part of our national heritage and a major tourist attraction.

DAVID STEPHEN

Contents

*Please note that the left-hand picture on page 50 is the Bank Vole, the Field Vole is on the right.

Red Deer

Tri aois coin, aois eich
Tri aois eich aois duine
Tri aois duine aois feidh
Tri aois feidh aois fir-eoin
Tri aois fir-eoin aois craoibhe-dharaich

Thrice the age of a dog, the age of a horse; thrice the age of a horse, the age of a man; thrice the age of a man, the age of a stag; thrice the age of a stag, the age of an eagle; thrice the age of an eagle, the age of an oak-tree.

Thus the old rhyme of the Gael. And in the Highlands today the ear can still listen to rhyme, ignoring research, and hear the great tales of olden times: when the proud, high-antlered harts were the quarry of kings; when King James V, with twelve thousand men, killed eighteen score harts in Teviotdale and thirty score in Athole, along with roe and roebuck, wolf and fox and wildcat; when two thousand Athole men, gathering the deer in Mar and Badenoch, Athole and Murray, could drive a herd, numbering a beast for every man of them, to delight the eye of that Mary of Scots who died at the hand of Elizabeth of England.

They tell yet, in the forests, of the milk-white hind of Loch Treig, that MacDonald of Tulloch knew: she who was never fired at and who lived for a hundred and sixty years in the wilds of Lochaber. In Badenoch roamed the Great Stag—the Damh Mor of the legend—who lived for two hundred years. Always there are stories of the great beasts of bygone days: who roamed the ancient forests through many reigns; who outlived the Chief and his sons; who were young when men had grown old. The kings and the chiefs have gone; only the deer remain. And the legends.

But the legends—thrice the age of a horse, the age of a man; thrice the age of a man, the age of a stag—are dying under the remorseless scrutiny of modern research; and the life of a stag can be measured by the rise and decline of his antlers, as a salmon's is measured by the scales, or a tree's by the rings of growth. At twelve, the Highland stag is in his prime—a Royal if ever he is going to be one; thereafter he begins to fail in prowess and antlers, and is old before men have attained to manhood.

The great Caledonian Forest that the Romans knew has gone; the deer forests of the twentieth century are treeless barrens, home of mountain fox and marten, wildcat and eagle, into which the new woods of spruce and larch and pine, planted by a new generation of men, creep slowly. The wild glens, the naked slopes and the high ridges are the territory of the red deer; true descendants of the great beasts of the Pleistocene: the largest and noblest land mammals surviving in Britain.

That is when the seals are all in the water. When the grey and common seals are ashore, he is the third biggest.

A Red Deer stag will stand 4 feet tall at the withers; the biggest ones go to 4½ feet. Size and weight vary greatly with feeding and geography, and there is no such animal as the average Highland stag. Highland heavy-weights are in the order of 20/21 stone. More usually, stags vary from 14 to 16, or up to 18 stone. Hinds are smaller and lighter, standing less than 3½ feet tall at the withers and weighing, as a rule, half the weight of the stags, age for age, on the same ground.

Only stags grow antlers, and each is known by the number of points or tines he carries—8-pointer, 9-pointer, and so on. A beast with brow, bay and tray tines, and three points on top, forming a cup, is a Royal. A 14-pointer is an Imperial. A beast with no points above the brow tines is called a switch. A hummel stag is one that has never grown antlers and never will. Occasionally, a hind will show horny rings or burrs on her head, so could be mistaken, on a casual glance, for a hummel.

The red deer society is a Matriarchy, and the sexes live in separate herds for most

Red Deer Stag in Hard Antler—after the rut at the beginning of winter. When the stag retires exhausted by the rut, he is said to be run-out. The run-out stag is in poor condition, but soon makes this up if the weather is favourable. But an early severe winter following close on the rut can cause some mortality among such weakened beasts.

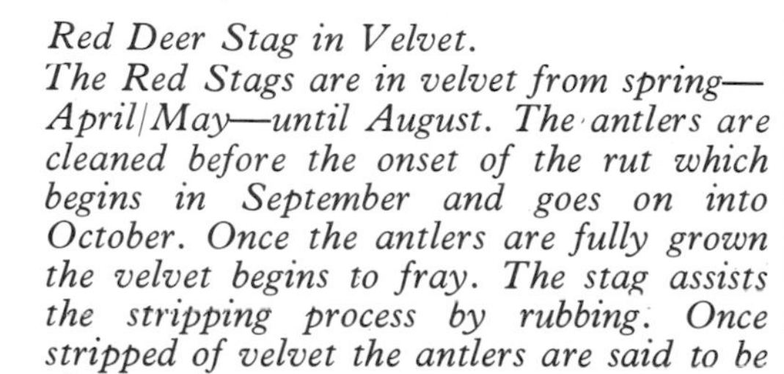

Red Deer Stag in Velvet.
The Red Stags are in velvet from spring—April/May—until August. The antlers are cleaned before the onset of the rut which begins in September and goes on into October. Once the antlers are fully grown the velvet begins to fray. The stag assists the stripping process by rubbing. Once stripped of velvet the antlers are said to be clean.

of the year. Hinds and stags keep to their traditional ground. From time to time, there is some overlapping but little real mixing.

Storms bring deer down from the high ground at any time. Even in midsummer, a freak storm will bring them right down to the glen bottoms. In winter, they will start to move well ahead of a blizzard and are seldom trapped in deep snow. When the snow lies deep, and the hills are tormented by blizzard after blizzard, the deer make long treks to lower ground. Then you will see great herds of stags and hinds coming down in long columns, high-stepping and creating their own storm of drifting snow with their hooves. In such situations, one can see as many as 500 deer in one group, sometimes of one sex, and sometimes mixed.

In a severe winter, many deer die and many hinds lose their calves. But the bodies don't lie frozen on the hill, to thaw and rot and pollute the air in spring. They are cleaned up by the scavenger force of foxes and eagles, ravens, buzzards and crows.

Hind ground remains hind ground, except during the rut, when the stags break in and take over. It is also the ground where the hinds drop their calves, and these can be found on the same terraces and flats year after year. But red deer are not aggressively territorial. Range would be nearer the truth of the matter because the deer cover a vast area of hill and glen, moving up and down through the contours according to the weather and time of year. Even during the rut, a stag holds hinds rather than ground.

The breeding season of red deer is the rut which is in September and October. The onslaught of the rut is heralded by the roaring of the stags when they break into the hind ground to collect their harems. Master stags collect as many hinds as they can and try to hold them against other stags coming fresh to the rut. Then the master looks what Landseer called him: Monarch of the Glen.

But his reign is brief, more apparent than real, because the hinds he holds are still a closely-knit social unit, following their own trusted leader. They accept the stag's overlordship for the duration, but are indifferent to it. In an emergency, the leading hind makes the move and her group follows. The stag has to follow on or lose them. The master stag cannot last out the long drain of the rut and, sooner or later, gives way to a fresher animal. After the rut, the heyday of the stags is over and the sexes re-group into more or less separate herds from then on.

In late May or June the hinds drop their calves, and are fully occupied with them while the stags are growing their new antlers. The red deer hind is an attentive mother and spends much time licking her calf and talking to it. Hind behaviour is highly social, and members of the group will combine to face fox or wildcat. Resting hinds will watch over the calves of others who are feeding some distance away. In summer, when the herds live high, a man can spend a pleasant half-waking night watching the hinds and listening to the small talk of the calves.

The stag calf produces his first antlers when he is about a year old, and cleans them of velvet in August or September, when he is fifteen or sixteen months old. Then he is known as a knobber. Antlers are cast and re-grown each year. The time of casting is March to May. During the growing period, the new antlers are covered in velvet which is rubbed off after they are fully grown.

Cast antlers are chewed and eaten by both sexes, but especially by hinds, presumably for the bone calcium. But the main food of Highland red deer is grass, heather, blaeberries, lichens and mosses, with the addition of browse when they are on lower ground. Deer within reach of the sea will eat a variety of seaweed. On low ground, they can do considerable damage, and marauding deer are sometimes a problem to farmers and graziers. The animals will browse wherever trees and shrubs are available, and they can do a lot of damage by stripping bark.

Outside the calf stage, the red deer has no enemies apart from man. Dogs can worry a calf, but this is a rare thing in the Highlands. Golden eagles take calves; so do foxes; but nobody knows how many they kill. During the period when the calves are vulnerable, mortality among them can be high; then the predators take them as carrion.

It is difficult to imagine a fox or wildcat killing a calf under the watchful eye of a hind. The golden eagle has greater advantages, but even the eagle isn't allowed easy access. The hind is likely to close with her calf and dab back at the eagle. I have known hinds dropping calves within a stone's throw of an eagle's eyrie, right under the bird's flight-line. There is no doubt at all, however, that the fox and the eagle kill calves. One finds the carcasses regularly in eyries and at fox dens.

In the Scottish Highlands, home of the big red deer battalions, control is carried out by stalking with the rifle, stags and hinds being killed selectively in proper season. The red deer population in the Highlands and Islands today is about 180,000 and this population is kept reasonably stable by the annual cull of about 30,000 head. In Scotland, the close season for red deer is from 21st October to 30th June for stags, and 16th February to 20th October for hinds.

Roe Deer

I sit among tall bracken on the face of the ridge, watching the sun go down in flames. The pines become ruddy in the glare; a raven croaks in the dark gully; two hoodies, flighting overhead, swerve and *Hah!* in surprise. Midges begin to attack.

Bough! Bough! Bough!

No dog-bark this; nor fox-yap; nor bark of adolescent roe—this is the *bough* of maturity, the chesty bass of a full roebuck, with the fires stoked for the rut, barking announcement and challenge on the night of 26th July, in a Highland glen, with a full moon rising.

At first light of morning I see him, mincing along the wood edge below me, with neck level and the spring in his legs. He has a fine head, with antlers heavily pearled and coronets clinkered. I watch him dingling fence wires, dirking posts, ruffling sallow, drawing low branches between his antlers—setting invisible scent markers as surely as a surveyor sets his pegs.

In another glen, on an August morning, I am perched on a high seat in a pine tree. On and off during the night a roe-buck has been rampaging in the thickets. Twice he has been under my tree, stabbing at the trunk. At daylight he is in a horseshoe of heather on the wood edge—an amphitheatre in the pines. He is parading.

With neck skewed and stiff he slow-paces, spring-footed. He furrows the peat with a forehoof, making a big scrape. He stabs the air sideways, knees down and pushes through the heather as though his antlers were a lawnmower. He rises again, takes three steps back, setting his weight on his hindquarters; then shakes his head, as though ready to launch to the assault.

The moment of action comes for him that morning when another buck enters the horseshoe. The newcomer is adult, but he has slender scrimshanks of antlers, with tines no longer than my eye tooth. They leap at each other like snakes striking; they stot back like dragonflies changing direction. Their hind-legs are springs launching their heads; they clash brow to brow with knees buckled.

They hit three times, then circle, the resident buck hooking round for the flank of the intruder. The pushing and hooking become a kind of dance. The fight takes them from the horseshoe, out of my sight but not out of earshot. I can hear the crashes of them. Presently the resident buck comes back: the victor. He is not pursuing beyond the boundary of his territory.

July 4th of another year. A roe doe rises from a rocky touzle of grass where bullocks are grazing. I cross to the spot. She has a fawn lying there, wet and unsuckled, still snuffling, with unshrivelled umbilical cord trailing on the grass like an earthworm. The doe circles me, barking, so I draw away. The buck appears, goes up to the fawn and sniffs it.

May 26th. That morning I am crossing the heather, yawny after a night's sitting, when a doe rises from a birch thicket with a spotted fawn sprackling in her slots. I turn away, not wanting to excite her, thinking she will have left a second fawn behind. She has. I see her with two the next morning.

An August morning out on the hill—a woolly kind of morning, with threat of rain. I am sitting below a crag, overlooking the wood. At the bottom of the slope, along the wood-edge, are several circular tracks—roe rings, trodden by the feet of deer. On one ring a buck is running a doe in anti-clockwise chase. This is mating ritual, roe-style.

Suddenly a yearling buck appears, nosing the ground, snorting and grunting. Not looking, but simply listening, a man might have thought for a moment that he was hearing a badger or a piglet.

The yearling has single antler spikes, mere finger-length dags. He bears down on the ringing buck, who leaves the doe to meet him. The master prods the yearling back uphill, rejoins the doe, and disappears into the wood with her. No argument; no duel; simple dismissal.

A full roebuck steps from the big timber into a new hill plantation. His antlers are rounded and clubby in the velvet, thick velvet like beaver lamb. The buck rubs his antlers against the stems of the young trees; he dances round while he does so. He frays the bark. He comes there day after day until the velvet is off and his antlers are clean. By then he has rubbed away the bark up to fraying height. The trees will now die.

These thumb-nails of roe are not just random tearings from a field notebook. They are chosen to illustrate important phases in the roe's annual cycle. They are particulars, and can now be related to the general.

The roe deer—slim of leg, cow-hocked, down in front like a hare—is the least of Highland deer: the small-antlered deer, the *Boc Earb* of the Gael. A three foot rule, stood on end, would top the withers of the tallest buck by six inches or more. A small buck may be little more than half the height of a Highland red deer stag.

Roe are also the lightweights among native deer, perhaps a quarter of the weight of a stag. Does are lighter than bucks. The best roe come from woodland, or other good habitats; beasts on the hill are leaner. Woodland is the roe's natural niche, but you'll find them on the hill sharing the spartan life of the red deer.

Territory size varies according to area, the smallest territories being in the best habitats. The master buck defends his ground against other bucks; not against a doe or does. But the resident doe will drive off an intruding doe. The buck marks his territory with scent from the glands between his antlers. He also frays vegetation, rubs trees, and makes scrapes in the ground with his forehooves.

The rut is from late July until the middle of August, give or take a few days at either end. Fawns are born the following May or June, but there are records for April and July.

The apparently long period of gestation is the result of delayed implantation until the end of the year, more than four months after the rut.

The doe leaves her fawns hidden for a few days until they are strong enough to follow at foot, returning at intervals to suckle them. She calls *whee-yoo* to them to rouse them; they answer with a bird-like peeping call. Adult roe grunt and snort. The dog-like bark can be a warning signal or merely to keep contact. A buck on rut barks more than usual.

A buck fawn cleans his first antlers before his first birthday and will take the rut, if allowed to, when he is 15 months old. Usually he isn't allowed to. But it can happen. I have seen it happen.

Pedicles are the young buck's first head-gear—furry protuberances on which the antlers will presently grow. The pedicles appear in August/September and are fully grown by November/December. Before the antlers appear the young buck grows horny tips on his pedicles, popularly known as buttons. These fall off in February, and antler growth begins on the raw sores.

The summer coat of the roe is foxy red; the winter hair is greyer. In winter coat roe of both sexes usually have two grey-white flashes on the gullet. The rump patch becomes whiter. The doe grows a white anal tuft, down-pointing, that looks like a tail. In fact, roe have no visible tails. The anal tuft marks the doe when the bucks are without antlers. The buck sheds his antlers in November/December and his new ones are complete by the spring.

Roe are browsers, and eat almost anything. On the hill they eat heather as well as grass. Juniper is an important winter food plant in some areas.

Roe are widely distributed in the Highlands, but density varies from one area to another. They are absent from the Outer Isles. There are a few on Islay, but none on Jura or Mull, Rhum or Skye.

Roe Buck in Hard Antler at the beginning of the rut in July. The Roe is the small-antlered deer. Roe antlers are usually a little over 9″ in length, although some grow up to 11½ inches. The antlers rarely have more than six points. Roe are in hard antler from March/April until November.

Sika Deer

The Sika Deer looks drawn in, down in front and behind, which gives it a skulking gait at the walk. Otherwise, its movements are very like the red deer's. Several races of Sika deer have been introduced to Britain but the one that has proved most viable in the wild state is the Japanese, and this is the type found in the Highlands.

The sika stag stands about the same height as a fallow buck, and the hind about the height of a fallow doe. The stag's antlers are lighter than the red deer's and less branched, usually carrying six points and not more than eight. The velvet is red with black tips. The first antlers are simple spikes as in the red deer knobber.

In its summer coat, the sika is as red as the roe; the winter coat is much darker. The calves, like the calves of the red deer, are spotted at birth. The sika's white rump patch is edged with black, and heart-shaped when spread in alarm. The tail is white.

But it is an easy matter to confuse the sika and the red deer. Although the sika is smaller, size doesn't mean very much in the field. Nor does coat colour, because many red deer are as red as the sika. One certain method of identification is the velvet. The red, black-tipped velvet of the sika identifies the stag at once. The antlers are in velvet from May to August.

The rut of the sika deer begins in September and goes on through October into November. Like red deer stags the sikas are polygamous, gathering small harems of half a dozen hinds or fewer. During the rut the sika stag whistles; a sound uttered by no other British deer. The stags will engage antlers and push, and sometimes they display real fire. But there is very little of the commotion so characteristic of the red deer stag at this time. The sika whistles and grunts, and maybe whistles and grunts again, and yet again; then he is silent for up to half an hour or even an hour.

Run-out stags, like run-out red deer stags, retire after the rut, and either live alone or in couples or trios. This depends on how many other deer are on the ground. Single stags or very small units are much more likely to be seen than large herds.

The stags tend to join up early in the year and then they run together until late spring, after they have cast their antlers. Outside the rutting season, it is quite usual to see sika grouped by sex as in red deer. But it is not uncommon to see mixed herds. However numerous sika deer may be on any ground, it is not usual to see them in large herds like red deer.

For much of the year, the hinds live in small units, perhaps up to a dozen animals. They hold this formation for most of the year, but hive off to give birth to their calves, and single hinds with a calf at foot, or with a calf and a yearling, are common enough sights in June and July.

During the day, sika usually lie up in cover and are not often seen on foot unless pushed out. They feed at dusk and during the night, and retire just before dawn. But in winter, when the snow is lying, they will often feed later on into the morning and come out earlier in the evening.

Sika deer are most likely to be found near woodland. They are beasts of the forest edge, where there is plenty of ground cover, and they like thickets of hazel, bracken and brambles. They will also be found lying up in deep heather, in clumps of unbrashed conifers, or in birch and alder swamps. On the hill, they eat the same food as the red deer and are extremely fond of the shoots and bark of the hazel.

Sika deer are found in Argyll, Caithness, Inverness-shire, Ross and Sutherland.

Fallow Deer

Many people, including myself, tend to look at the fallow deer over the end of their noses, considering it a half-tame park animal. But where a herd has long been feral, the fallow can be as wild as the red deer or the roe, and just as difficult to get to grips with.

Fallow living wild behave differently from fallow in the sanctuary of a Deer Park. But they don't all behave in the same way. Much depends on local conditions, pressures and upsets.

Fallow deer in a park are there to be looked at, are used to being looked at, and can usually be seen at any time of the day. Feral deer can also be seen by day, moving, resting or feeding, where they are not being disturbed or harassed. But when they are under pressure, they become as shy and secretive as any deer, using cover as skilfully as the roe.

Although fallow deer will sometimes be found living in the open with red deer, sharing the same ground, they prefer deciduous and mixed woodland with plenty of ground cover and thickets. On more open terrain they will lie up in bracken and here their fawns are regularly dropped. Finding fallow in thick woodland can be difficult; stalking them can be even more difficult. A wily old buck will sometimes lie close in cover during daylight, feeding at nightfall and dawn.

The fallow buck comes to the rut early in October and the excitement is over by the end of the month or early in November. Like the roe-deer the fallow buck makes scrapes into which he urinates and which he rakes with his antlers. Each master defends his rutting stand against all-comers but serious fights are rare. The buck sets out his territory markers by fraying the bark of trees and bushes, and sets his scent at the same time.

There is a hierarchical system in this species, an order of precedence among bucks that is observed with some punctilio until the younger beasts feel strong enough to challenge the order. Then there will be a fight. But most so-called fights are displays of threat, and threat is usually enough to make the disputing buck submit. When a real challenger comes along the master may be toppled from his position after a severe fight.

After the rut herding is mainly by sex. Wild herds are usually much smaller than those in parks. Bucks of all ages mix but the older beasts tend to be solitary for much of the time. The does leave the herd to give birth to their fawns, then come together again; but it is difficult to fit a pattern to this. The fawns are playful and the does spend a great deal of time with them when they are small, skipping this way and that, running in circles and bounding stiff-legged to double their own height.

A mature fallow buck stands under or over three feet at the withers. A very good wild buck might weigh fifteen stone. Does are smaller and lighter than bucks.

Although there are many colour varieties in this species, there are two main phases: a dark type with hardly any visible spotting, and a light type with obvious spots. The tail is black with white on the underside. Fawns are spotted at birth—with white in the light varieties, indistinctly in the dark ones. Only the bucks grow antlers, which are palmate, broadening with maturity, and at their best in the seventh to ninth years. The antlers are cast in April or May and the new ones are clean by late August or September.

Fallow Deer can be seen in varying numbers, but mostly in small units, in Argyll, Banff, Caithness, Inverness, Ross and Cromarty and Sutherland, and on the Islands of Islay, Mull and Scarba.

Reindeer

The native reindeer of the Scottish Highlands became extinct in the twelfth century. Today's Highland reindeer, found only in the Cairngorms, are the descendants of Swedish stock re-introduced in 1952.

Professor Ritchie blames the disappearance of the native reindeer on the progressive destruction of the old forest, although the last animals were probably exterminated by man, for it is known that they were being hunted to the end by the Jarls of Orkney in Caithness.

Many centuries passed before any attempt was made to re-introduce this lost species. In the eighteenth century, fourteen reindeer were brought to Dunkeld by the Duke of Atholl. Only one of these animals survived, and for not more than two years. Other attempts were made to settle reindeer in the Forest of Mar and in the Orkney Islands, but both were unsuccessful. Successful re-introduction had to wait for the arrival of Mikel Utsi with his Swedish animals in 1952.

The Rochiemurchus herd lives a free life and exists almost entirely on its own merits, so cannot be compared with animals reared in zoos. The Swedish animals brought by Utsi have provided the foundation stock for the establishment of reindeer in other parts of Britain.

Reindeer, in North America known as Caribou, can be separated into two types —mountain and woodland. According to Ritchie, the Scottish reindeer were more woodland type than barren-ground type, despite the fact that their antlers did not make the size of woodland caribou's. The first animals brought by Mikel Utsi were of the mountain type. Later, the woodland type was introduced.

The reindeer were first released in June 1952 on 300 acres of land lent by Colonel J. P. Grant of Rochiemurchus. Reindeer of the forest type came later and, in 1961, a few animals were brought from Northern Norway. Since then, the herd has been breeding successfully and many calves have been born. In 1972, 31 calves were born, 21 of which have survived.

The Rochiemurchus herd is now well-known and a great attraction to tourists, although the animals were originally introduced as an experiment to discover if they could live and breed in a country formerly occupied by their species.

Why are there reindeer in Scotland at all? Mikel Utsi, Managing Director of the Reindeer Company Limited and Technical Adviser to the Reindeer Council of the United Kingdom, has said:
"They are useful in many ways. The meat is delicious, because the animal is bred for meat and can be slaughtered at the right time. In 1951–1952 about 400 tons of reindeer meat sold quickly in London. When available at Glenmore today, by order, it fetches a good price. Clothing made of the skin or well-tanned hide is worn even in temperate climates. Reindeer skins, simply stretched and dried, make warm groundsheets and are often bought as floor rugs in Britain. Reindeer hair has been imported into Britain for expensive dress materials. Antlers are carved for crafts. Trained reindeer oxen can pull goods or people on sleds, or carry packs, and children may ride them, too. Research into the composition of the cream-like reindeer milk, the digestion, skin, antlers, breeding and other research topics have been aided by the Reindeer Company's archives and the headquarters at Reindeer House."

Reindeer in North America and Scandinavia feed mainly on a lichen known as reindeer moss. They eat this lichen in the Highlands, too, although they take other food as well. Male reindeer are known as bulls; females as cows. Calves are born in May and June, and have unspotted coats of grey, brown or near-white. This is the only deer in which both sexes have antlers.

The Roe Deer doe can be distinguished from the buck at all times by her lack of antlers. In the winter, after the buck has cast his antlers, the doe can still be distinguished because of the tuft of hair below her rump, which looks like a tail. In winter coat, the roe has a white or grey patch on its gullet and another on its throat.

Left

Fallow Deer Buck showing palmated antlers. There are two colour varieties of fallow in Scotland—the dark and the menil.

Right

Blizzards and deep snow bring deer down to the low ground. These stags have come from the high ground to the river valley where they scrape through the snow to find food.

Below

Red Deer Stag photographed in Knoydart. This stag is an Imperial—that is a fourteen pointer—and has just come to the rut.

Left

Red Deer Calf about one day old. The hind leaves her calf unattended for long periods but always returns to nurse it. She is a very attentive mother. Hinds tend to cooperate in looking after calves. At this stage, the calf is within the prey range of the mountain fox and the golden eagle.

Below

Roe Deer Buck at the age of about eight months, showing his first antlers beginning to erupt. Roe deer are in velvet from late November until March or April. Adult bucks are six pointers, almost invariably.

Rabbit

Everybody knows the rabbit. Before myxomatosis, when big battalions were the order of the day, everyone was familiar with it. Yet it is a recent arrival; a nineteenth-century Highlander. Many people brought it to many places. According to Osgood Mackenzie, it was being planted at Gareloch around 1850. The rabbit plantation of the Highlands was as real a historical event as the Scots plantation of Ulster.

It is difficult to appreciate that the men who marched to join Charles Edward at Glenfinnan had never seen a rabbit, but it is likely that some of them knew the wolf.

Official policy today is to keep the rabbit down and, if possible, kill it out. The first may be possible; the second is a dream. It has survived predation by man, foxes, badgers, stoats, weasels, cats (wild and tame), dogs, hedgehogs, rats, crows, hawks, owls, buzzards, eagles. It has survived myxomatosis.

In wild rabbit populations, something like two-thirds of litters conceived never see the light of day. The embryos die about the twelfth day of pregnancy and are quickly resorbed in the uterus. This may be a form of built-in birth control.

The main breeding season of the rabbit is from January to midsummer, which means that the colonies are at their peak numbers in early autumn. But there is sporadic breeding in all other months.

Another phenomenon in the rabbit, that has been discovered, forgotten about, and re-discovered several times, is refection or re-ingestion. The rabbit swallows the pellets that pass through its bowel by day and makes a second meal of them. It shares this coprophagy with the hare. If you keep rabbits, wild or tame, you can observe it for yourself.

There is plenty of reference to it in the Old Testament, which probably had it first. Leviticus II, Verse 5, reads "And the coney, because he cheweth the cud, but divideth not the hoof; he is unclean to you". The same is said of the hare in the next verse. And of the pig in verse 7, although in reverse, because the pig has the cloven hoof but doesn't chew the cud.

The rabbit is a burrower and, where warrens are big, with great tunnel systems and scattered earth works, the area begins to look like a Flanders battlefield—scarred, upheaved and cratered. Big warrens, like a chain of fortresses, were traditional; but myxomatosis changed all that. It cleaned out areas that had been occupied for human generations. Small warrens come and go, but some develop into bigger and more permanent systems.

In each colony there is a hierarchical system—a peck order that operates above ground, but which appears to break down below ground. Old buck rabbits, usually heavyweights, are the masters of the colony and lesser bucks always give way to them. Dominant does, presumably the favoured mates of the master bucks, behave similarly towards lesser females.

In a wild population, it is impossible to guess at what happens below ground but in his artificial warren, built to give a round-the-clock view, R. M. Lockley noticed that there was perfect amity among the animals of the colony whatever their sex or age. My own observations of a control group, which I had for two years, confirm this.

Many doe rabbits have their young in the main warren, which is perhaps the safest place, so it is likely that they are the favoured mates of the master bucks. It is possible that the does that litter away from the warren, in special nursery burrows or stops, are lesser lights in the colony, although this is by no means certain.

The behaviour of a doe rabbit who has young in a nursery stop is largely predictable. She will not usually visit them in daylight but, in June when the nights are short, she will do so well before dusk and sometimes remain with them well into the forenoon. I have watched a doe, in June, close a stop at 11 a.m.

The rabbit is the "coinean" of the Gael, coney being the English of it. Although the rabbit breeds with such haste and fecundity, it has a built-in population brake in the phenomenon of resorption. A percentage of litters conceived—sometimes a high percentage—is never born. After a period of development, the young die in the uterus and are re-absorbed into the mother's tissues. Like hares, rabbits refect, which means that food passing through the bowel is re-swallowed and re-digested.

Sometimes a doe rabbit will make a nest above ground in a grass tussock or under a bush. In a case like this, she has to open and re-make the nest each evening when she comes to nurse her family. A nest like this is very likely to be destroyed by fox, hedgehog, or badger.

After being deserted by their mother, at about the age of four weeks, the young rabbits may remain near the stop for a time or move at once. If they are roughly treated at the warren, they will live on the fringe of it, either holed-up or lying out, or they may form a small colony of their own in a hedge-bottom or bank nearby.

Rabbits have a regular, well-padded system of runways leading from their burrows to their feeding areas. The runways are most obvious near the burrows. From a warren in a wood, the runways will pass through gaps in netting, holes under an old wall, or breaks in a wall, or under the exposed roots of a tree.

When pursued or frightened, the rabbit, homing on his own runway, makes haste with confidence, knowing the ground ahead. Put off his route, he begins to dodge and panic. Although he goes off with a great burst of speed, he is not a distance runner; if he doesn't quickly reach his burrow, he can be caught by a fox or dog in the open, sometimes even by a man. Many poachers have the knack of stalking and picking up rabbits in heavy ground cover.

Rabbits emerging to feed come out cautiously, with nostrils twitching and ears wagging, and I have watched a general retirement when a group came up against the scent line of a fox that had passed that way earlier. The taint of a fox lying about will keep rabbits in and they come out most circumspectly once it has lifted.

Although it is the bread and butter of any hunter fit to catch and kill it, the rabbit sometimes shares a burrow system with its predators: fox, badger, stoat, cat or rat. In other parts of the Highlands, it will share its burrows with sea birds. I have seen rabbits living with all of them. On the Island of Noss in Shetland, which is a National Nature Reserve, I have seen rabbits, cats and Shetland Lambs come out of the same burrow.

But this sharing of burrows is often no more than a next-door arrangement, rather than cohabitation. The rabbits stay in one part and the real owners in another. Once I saw fox and rabbit using holes only four feet apart. I dug in several feet and found that the tunnels diverged widely. After the fox has scraped out as much as she wants, the rabbits are safe so long as they don't enter her section, which they are not likely to do.

For two seasons I had a vixen with cubs under observation in an extensive rabbit warren. This was long before the arrival of myxomatosis, and rabbits were thick on the ground. I often saw rabbits sitting around the burrows, or fanning out into the field, when the vixen was coming home at first light, but she made no move to molest them, even when she was carrying nothing. The rabbits used to sit up at her approach, or those nearest her would draw aside a bit to leave her a wider clearway.

My only experience of feral cats sharing quarters with rabbits was on the Island of Noss. All over the island, I found bits and pieces of young rabbits that had been killed by the cats; but at one section of the warren, where I watched a cat for a week, I counted the same number of rabbits alive each night over the period. The cat had been killing away from home.

A stoat living in a warren has the advantage of being able to enter every corner used by the rabbits. I have seen a stoat actually play with rabbits belonging to the warren in which it was living. The play was never rough or threatening. In fact, it looked more like a neighbourly pat on the back than anything else.

The rabbit is found all over the Highland mainland and on most of the islands, but it is not found on Rhum, St. Kilda or North Rona.

Blue Hare

Although the sky is like blue glass, with no snow falling, the icy wind raises a snow-storm on the slope, blasting jets and clouds of white uphill, and obliterating the wide front of hare tracks leading to the crest of the ridge. Finely spun snow creeps in waves along the slope, like a spent tide lapping. But on the crest the wind sweeps it out and up to meet other drifting clouds and merge into a swirling chaos that blots out the sun.

Down along the lea slope, where the snow lies hardly ruffled, the white hares sit out the storm, spaced out in small groups, but squatting singly in hollows, among boulders, or under peat overhangs. They sit under the fall-out from the storm high above them.

This is hare territory, from top to bottom of the ridge on both sides. The hares move up and down, and over the top, according to wind, weather, food, or fancy. Today they have come over on the wings of the storm, after sitting with their backs to it on the other side until their white fur was further furred with snow.

After midday, the wind changes the direction of its assault, and rakes both sides of the ridge. The blown snow stings like sharp sand. The hares sit on while their tracks are blotted out, and the snow tops up the hollows and piles great drifts against boulders and outcrops of rock. Then, singly, or in two and threes, they begin to move down, spectral shapes in the snow swirl. Far below, they break out of the storm into the open sunshine, and now each hare skips over the drifts, and the sparking flats, on snow-shoe feet, accompanied by its bobbing shadow. A man watching from the river below could have counted sixty of them moving out of the storm, not in a flock like sheep, or even deer, but in singles, twos and threes, in extended line.

The storm lifts before sunset and, in the stillness, the first stars presage frost. The hares begin to move, nibbling at heather cleared by the deer, and on frosted cotton grass. Then, at darkening, they begin to move back uphill, each returning on its downhill line. At daylight, they are back in their seats near the top of the ridge.

Mountain hares move up and down like this, according to the vagaries of the weather, and will change their seats to escape wind and storm. Blizzards and sudden severe snow-storms will drive them down from the heights, and a really big snow will keep them low for a day or two; but they move up as soon as conditions are suitable, and then their white shapes can be seen grazing on dark patches where the wind has cleared the ground. Movement is not predictable. A light snowfall, even when followed by high winds and drifting doesn't always move them, because they can lie up in natural holes or in burrows they dig in the peat. Even when these burrows are drifted over, they can get through to them by tunnelling in the snow. In such places, the hares lie up, snug and safe, until the storm blows over.

But a lasting blanket of snow, with prolonged blizzards can put them down and keep them down until the higher feeding areas have been blown clear. In the worst weather, they will feed in the glen bottoms or in woodland, eating heather, rushes, bog-cotton, or the bark of trees and scrub. When hard pressed by hunger, they will even eat pine cones. At such times, great gatherings are sometimes encountered, and I once met such a mass exodus during a severe blizzard, when the column of hares took several minutes to cross the road in the car headlamps.

Look you, now, at a gathering of hares in the mist and frost of a March morning. Below the snow-line, on a boulder flat where the drifts are islets of white in the thaw, eleven hares are assembling like blackcocks coming to a lek. They are in the moult, with dusky fur ousting the white on face, neck and shoulders. Tufts of white fur cling to heather twigs and lichened boulders, like blackcock feathers on a lek.

The hares move in until they are occupying about half an acre of ground. Three are hopping slowly on a short circuit. Two are padding over snow-drifts, where the others crouch or sit beside boulders or heather-clumps. The upright hares have their black-tipped ears pointing to the sky. At this point, everything is casual, like lekking blackcocks relaxed with their tails folded down.

Suddenly, the hares come to life, as blackcocks will explode suddenly into activity. They leap and kick, and run in circles and figures of eight, chasing and counter-chasing until they are covering well over an acre of ground. They skip over the drifts, kicking up clouds of snow. They kick and box, retreat and box again. Fluffs of fur float in the air or snag up in the heather. The aggressive display and the frolics end when the hares have run wide enough to be out of contact. Then they begin to drift away, singly, like blackcocks walking off a lek before taking flight.

After mating, the doe hare goes her own way until she is ready to come to the buck again. When her young are due, she scrapes out a form for them in the heather, or she may give birth to them in a peat shelter or burrow. From the beginning, the leverets are active, being born furred and with their eyes open. They squat in their forms, but if put on foot by dog or fox, or disturbed by man or bird of prey, they will bolt below ground to safety. I once saw a small leveret come out of a peat burrow where wheatears were feeding young. I have found twin leverets at over 3,000 feet in the third week of March. Adam Watson has found leverets at over 4,000 feet.

The blue hare is the true hare of the mountains, found usually from 1,000 to 2,500 feet above sea level, although it will range up to the 4,000 feet contour. It will also be found at sea level. In many places, its range overlaps that of the brown hare, especially at the 1,000 to 1,500 feet contours.

In summer coat, the blue hare might easily be mistaken for the brown, although it is a smaller animal—more thick-set, with a bigger head, shorter legs and shorter ears. But these are not the sort of characteristics that are obvious at a glance. The real difference is in the tail. The tail of the blue hare, at all times, is white above and below. In winter there is no difficulty, for the mountain hare is the only British mammal, apart from the stoat, that changes to white.

The summer coat is dusky or deep brown, frosted or grizzled, with a bluish bloom. The ear tips are black. The change from brown to white, as in the stoat, is the result of a moult. First of all, in autumn, the hare moults into a new brown coat; then the white coat begins to appear, sometimes as early as October. Usually the moult is complete by mid-November or early December. The change begins slowly and is completed quickly. The summer coat begins to grow in again between March and early May.

Blue hares feed mostly in the evening and during the night, but in winter they will feed at any time. When the snow is lying deep, they can be seen on foot and feeding at all hours of the day. Re-ingestion takes place by day. The mountain hare, like the brown hare and the rabbit, re-swallows its soft daytime droppings, which are then passed through the digestive system a second time.

The mountain hare is neither so fast nor so strong or resolute as the brown; nor does it appear to have any of the brown hare's tricks. A good hill collie can run one down in the open and the hare seems to have neither the strength nor the will to make a hard run of it. It will sit close at the approach of man or dog, sometimes until almost trodden upon, then break away uphill; but it probably won't run far before stopping to sit up and take a backward look.

Apart from man and his dogs, the hare has most to fear from the mountain fox, the golden eagle and the wildcat. The mountain fox will track a hare, course a hare, or dig out a hare. One day, during prolonged snow, I followed mountain fox and hare tracks along the 1,500 feet contour and came on blood and hare fur half a mile

Mountain Hare in winter coat.
The mountain hare is one of the two British mammals that turns white in winter: the other is the stoat. Both are found on the same ground. The hare in winter white is difficult or impossible to see against a background of snow; when there is no snow it is conspicuous.

later. The fox, of course, follows such tracks by the scent lying in them, and not visually, as a man does.

The mountain hare of the Highlands has been introduced to many parts of Scotland, England and Wales, and is now found on some Hebridean Islands, Shetland and on Hoy in Orkney. Irish hares are a closely related species that have been introduced to the Island of Mull.

Brown Hare

The hare is buck-toothed and big-whiskered, with long hind legs, long black-tipped ears and a short tail that is black on the upper side and white underneath. The general colour of the fur is reddish-brown or greyish-brown, but there is some variation with the season and the amount of wear on the hairs. The winter coat is thicker than the summer one. Leverets, at birth, have a wavy coat of reddish-brown, streaked and flecked with black and silver.

Generally speaking, hares like open, undulating country—up to two thousand feet and down to sea level. Most are found below a thousand feet—on marginal farms, moorland, dunes, and even in woodland.

Speed and endurance, not to mention courage and a high degree of cunning, are the hare's defences in the fight for survival. She is built for speed and is at her best uphill, which is the way she usually runs. When turned about, she will run a diagonal rather than a straight descent, for her long hind legs put her at a disadvantage on a fast downhill run. A fox will sometimes course a big hare in the open but prefers to chase younger animals lacking the speed and endurance of the adults. But a very persistent fox may track a hare for miles.

Depending on its size and condition, the hare is preyed upon by foxes, eagles and wildcats; domestic cats, dogs, stoats; polecats, ravens and buzzards. Foxes can and do kill adults. So do wildcats and eagles, but the number they kill is another matter. Most hares are killed as leverets by wild predators. Small leverets are within the prey range of the stoat. Only the tiniest and the most helpless come within the prey range of the weasel. Many farm cats can kill leverets more than half grown, and up to a weight of six pounds. Crows, ravens and buzzards usually take them small. A fast sheepdog can be a considerable predator on leverets, but man is the most persistent and deadly predator on the brown hare.

In March, brown hares traditionally go mad, but the madness really begins in February and is at its peak when the rooks are nesting. At this time, these otherwise solitary animals hold their gatherings, like blackcocks coming to a lek. All the hares in the neighbourhood move to the trysting place—in a field or on a hillside, or some open space in a gorse or thorn-brake. There, they chase each other and leap, and kick, and box, and run in circles, and behave generally in a wild and boisterous manner.

Bucks kick bucks, and does kick bucks, but for different reasons. The bucks compete with each other, whereas a doe fights off suitors pressing her before she is in breeding condition. Sometimes the fur flies, but serious fighting is rare. A buck with nose to ground will run with a doe in circles and loops until he is challenged by another whom he has to box. The harder hitter or kicker takes up the running with the doe hopping or spurting ahead. Sometimes there is a general flare-up, as you get with blackcocks on a lek, or a pair will hive off and, for a day or two, a buck can be seen trailing after a doe along hedgerows, across fields, over ridges, or through a wood. Pairs of hares are quite a common sight after the general assembly, the bucks dawdling after the does everywhere they go.

It is strange that the hare, which breeds three or four times a year, should have this spectacular burst of activity in early spring and nothing afterwards except the discreet pursuit of a doe by a buck along quiet trails. Why do hares gather like this? How do they know where to meet?

There is no doubt at all that these trysting places are traditional and the assumption is that the hares know where to come, and that they come from far-flung territories, where each has been living a solitary life for months. Thus, the sexes can meet and mate. Apart from this spring trysting period, hares lead solitary lives. The adult does, of course, have a close association with each of their several families, but the association lasts for little more than a month and the break, when it comes, is final and decisive.

Leverets are born above ground, fully furred, with their eyes open, and are able to run about shortly after birth. They are born in a pressed-out hollow in the grass, or in a tussock, or under a hedge, or in a wood. The nest is known as a form. Very soon after birth, the leverets separate, each finding a little form of its own. The doe returns at intervals to suckle them, usually at night, when they all run to her. The young hares are self-supporting about the age of a month.

The place where a hare lives, where she lies up by day, where she rests and sleeps, is called her form or seat, and she will have several of these on her range. She may use one for a fortnight or more, but she is sure to move from one to the other according to the amount of disturbance, the direction of wind, the weather, and the temperature. In hot weather, she likes a form that gives shade. In cool weather, she likes one that catches the sun, and all hares like protection from the prevailing wind. The form may be a warm nest pressed out in a thick grass tussock, a hollow in a field with only a few blades on either side, a hollow beside a sod or hummock, or a pressed-out seat beneath a gorse bush. Woodland hares often have their forms in thick cover, with well trodden runways leading to the fields, and they will bite off intruding vegetation to keep these lanes clear.

During most of the day, the hare rests in her form, sleeping, or watching, or re-ingesting. Like the rabbit, she re-swallows her droppings when they first pass through the bowel. In the evening, she moves out to feed. Because the hare is always so alert, going away at the slightest sound or taint on the wind, she is difficult to surprise in her form by the ordinary stealthy approach, and this has given rise to the legend that she sleeps with her eyes open. When returning to her form, the hare has a ritual of caution that is inborn. She takes great leaps sideways before she moves in, thus breaking her scent line. This makes it difficult for a predator, like a dog careering along on a hot-line, from running on top of her, and gives her time to move while her trail is being unravelled. When she is ready to leave her form, she does no trail-faking or trail-breaking. She rises, and goes.

Hares swim well and readily. They don't drink much except in very hot weather, when the vegetation is dry. Tame hares drink at any time, and I have seen a wild one standing up against a cattle trough to reach the water.

When the snow is lying deep, the hare is able to leap from a drift over the netting into a planting, and jump her way out again. In hard weather, she will visit farm stackyards at night to eat hay or turnips, or sneak in for a bite where out-wintering bullocks or ponies are being fed on the hill. I knew one that lay up for three days in a snow-drift following a blizzard. In this way, the hare conserves energy. It is better to lie doing nothing in such conditions than expend energy trying to find food.

A hare on foot by day will generally clap down as soon as a man or dog comes within sight and keep on flattening while a closer approach is made. Often, such a crouching hare is mistaken for a lump of brown earth, and so a lump of brown earth is often mistaken for a crouching hare. There is an old saying, and a true one, that covers this. If the object you see gets bigger as you approach, it isn't a hare. If it gets smaller, it is a hare.

Hedgehog

It is a truism that mammals are far more difficult to work with than birds—shy, elusive, wary, less easy to watch, less easy to come to grips with. This is not true of the hedgehog. No mammal is easier to meet, look at, keep in touch with, or follow on his rounds. You can stay with a hedgehog for practically the whole night if you don't crowd him too closely. You don't even have to get down on your hands and knees, although this can have its advantages.

If you come right up to him he will lower his brow quills and crouch, waiting for your next move. If you touch him he will ball up, and you are left with a hedgehog defence you can do nothing with. But he won't stay like that for long, if you step back again and let him go on his way.

Unless he comes on a big prey, like a frog, or a vole, or a nest of rabbits, you won't often be able to tell what he is eating; but you'll know when, because he makes plenty of noise about it.

He eats a great variety of invertebrate prey, including beetles, earthworms, caterpillars, snails and wood-lice. He eats slow worms. He takes some vegetable matter—soft fruit, berries, even acorns at times; but it is doubtful if this amounts to much. Nestlings and eggs make up only a small part of his diet, although you'll hear it said that he is a great predator on the eggs of game birds.

The fact is that he is a beast of catholic tastes, and takes what he finds; he doesn't go looking for anything in particular, except when there is a seasonal abundance of something—like young frogs in late summer, or a lot of earthworms after rain. He takes what he can catch and hold, so his prey range includes, at times, mice, voles, lizards, frogs, small rabbits, and rats. He will also take, if you make them available to him, the scrapings of the frying pan, pig meal, poultry pellets, and kipper skins.

The hedgehog is a true hibernator, and puts on fat for the long slow-burn of his winter sleep. During hibernation he gives up his temperature control; his heart and respiration slow down. He gets as near looking dead as it is possible for any animal to get without actually dying.

For his winter sleep the hedgehog likes a warm bed, insulated against the cold. The bedding he gathers becomes matted into his quills. The hibernation period is from October to March, depending on weather conditions. I had a pair that slept without a break from mid-November until the last day of March. During the

Hedgehog and Adder
It is an established fact that hedgehogs can and do kill adders and eat them. It is also a fact that some hedgehogs will retreat from an adder. When such a confrontation takes place, the snake strikes at the hedgehog, which squats with brow quills down to guard its face. Adders cut themselves on the hedgehog's quills and bleed. At this point the hedgehog either moves in for the kill, biting through the snake's spine, or it runs away. It is easy enough to understand why a hedgehog should attack an adder. It is much more difficult to understand why the snake should attack and keep on attacking until it lacerates itself.

long sleep the male had lost 10 ounces in weight and the female 7 ounces.

After waking the hedgehog comes to gradually, and for a week or so afterwards is liable to fall asleep at any moment, or lie up for a night. I once found one fast asleep with an earthworm wriggling from the chink in his armour, and another dreaming on a frosty night with a live frog in his jaws. A hedgehog that has caught a big frog will sometimes start eating without taking the preliminary step of killing it.

Boar hedgehogs fight with each other in spring, presumably in defence of territory, and if one marks a male one will almost certainly find him on the same ground night after night, and associating with the same female and her young. When the sow has very small young in the nest she will keep the boar away, but as they grow older there is free association. This is the general picture, borne out by the behaviour of hedgehogs in captivity.

A sow hedgehog with small young returns several times in the night to suckle them, and lies up with them by day. If the breeding nest is much disturbed, or the young are handled a lot while still very small, the sow will move them to a new nest, or even desert them. When moving young she carries them like a cat carrying kittens, but some females make do with a leg-hold, which results in squeaks of protest from the squirming victim.

The hedgehog is notable as a killer of adders. There is no doubt at all that hedgehogs do this, and eat the snake. I have been unlucky in not having seen this, although I have seen a few confrontations. In Argyll I saw a hedgehog retreat from an attacking adder. Two hedgehogs that tried to kill adders were bitten on the nose and died the following day. The hedgehog is not immune to adder venom.

It used to be believed that hedgehogs sucked cows; there are people who still believe it. There is no truth in the belief, but there is an association between hedgehogs and dairy cows at pasture during the night. The hedgehog snuffles about where a cow has been lying, and sometimes finds globules of milk on the pleated grass—the squeeze-out from heavily stocked udders. The hedgehog's mouth could never span the teat of a cow, and no self-respecting cow would stand for even a first attempt.

The hedgehog's quills, and his ability to roll himself into a tight ball, are his defence against his enemies: foxes, badgers, dogs. Some dogs can open hedgehogs; some foxes can do it; and any badger can do it. The badger simply tears the beast open with his bear-like claws. The fox gets a tusk into the chink in the armour, where the hedgehog's nose is tucked into his feet. An early hedgehog will sometimes be attacked by a waukrife crow.

Hedgehogs are not likely to be found in tall, closed coniferous forest, on very high ground, very wet ground, or heather moorland. They like the forest edge, hedgerows, open woodland, farmlands, rough pastures and orchards. They hunt from dusk until dawn. They swim well and readily, and will go into the water after frogs. They climb indifferently and fall often. They fall into water barrels and are drowned. They will walk into the same trap, or the same snare, night after night, and even within the hour.

In the Highlands the hedgehog is thinly spread, and local. It is found on the islands of Orkney, Shetland, Skye, Mull, Coll and Canna, to some of which it was introduced.

Winter quarters are chosen with some care; they have to be proof against frost, heavy rain and snow. Favoured sites are old rabbit burrows, a hole in a drystane dyke, under the exposed roots of a tree, in the bottom of a corn rick. The old breeding stop of a doe rabbit makes a warm hibernaculum. Where ground cover is thick, as in a bank layered with brambles, the hedgehog will be happy with a mere cubby hole. He carries the bedding in his mouth, then burrows in and pads himself thickly. If he is in a rabbit burrow he fills the entrance with leaves.

Right

The Japanese Sika Stag is as red as the red deer in Summer, but darker in Winter. His antlers are small, with at most eight points. The rump patch or caudal disc is white, and the short tail is partly hidden by the white hair. Roe deer in Winter have a white rump patch and this leads to confusion between the two. Sika can also be confused with dark fallow deer. The stag illustrated is the black variety of Sika, not found in Scotland. The Japanese sub-species (Cervus n. nippon) is now found in nine Scottish counties.

Below

The Soay Sheep is now found in small unit herds on many parts of the mainland, including the Wildlife Park at Kincraig near Aviemore.

Left

During the Pleistocene the Reindeer was a common animal in Scotland. It is often said that it disappeared in the 9th Century. It is more likely that it remained until the 12th Century because, according to the Orkneyinga Saga, the Jarls of Orkney were hunting it then in Caithness. Professor Ritchie gives his blessing to the latter view, arguing that the Scandinavian sagaman knew as well as anyone the difference between a red deer and a reindeer.

Right

The Hedgehog, although an insectivore, can be more accurately described as omnivorous. He will eat anything he can catch and hold. The hedgehog can climb rough walls, wire netting and trees, and can fall some distance without injury because his quills act as shock absorbers.

Below

Wild Goat Nanny with Kid. The kid has been licked dry and is only a few hours old. Single births are the rule.

Fox

Left

The Rabbit population now has spectacular ups and downs due to the ravages of myxomatosis. Before the advent of this disease, the rabbit was able to withstand the pressure of the whole predator force in Scotland. Now its numbers are periodically decimated and local populations are almost wiped out. There is some evidence, however, that the survival rate is becoming increasingly higher.

Far left

Young Rabbits in their nest of hay and wool. The question is often asked: what is the name of a baby rabbit? The answer is that the adult is, strictly speaking, a coney. The young coney is a rabbit, formerly rabet. Nowadays we call adults rabbits, while young are often referred to as kittens.

Thou shalt not suffer a fox to live! might well have been the Eleventh Commandment if Moses had come down out of Cruachan instead of Sinai. There is no ambivalence in attitudes towards the fox in the Highlands—no inconstant love-hate relationship or sneaking admiration for the animal of folklore or legend. It is the most unequivocally hated and anathematised mammal on the Highland list.

Yet it was not always so. Somewhere between 1790 and 1804 Duncan Ban MacIntyre, that very great Gaelic poet, wrote his *Oran nam Balgairean*—the Song of the Foxes. It was a bitter comment on the coming of the sheep, which made him popular in his day but would hardly make him popular now:

My blessing be upon the foxes, for that they hunt the sheep—
The sheep with the brockit faces that have made confusion in all the world,
Turning our country to desert and putting up the rents of our lands.
Now is no place left for the farmer—his livelihood is gone;
Hard necessity drives him to forsake the home of his fathers.
The townships and the shielings, where once hospitality dwelt,
They are now nought but ruins, and there is no cultivation in the fields.
Deeply do I hate the man who abuses the foxes,
Setting a dog to hunt them, shooting at them with small shot.
The cubs, if they had what I wish them, short lives were not their care.
Good luck to them, say I, and may they never die but of old age.

An ultra-modern comment comes from Sir. F. Fraser Darling and Dr. Morton Boyd in their revised *HIGHLANDS AND ISLANDS*: "Highland sheep farming has been one of the depressants of biological condition of the wild lands, and the intense persecution of the fox, which does so much else than eat lambs, has been one of the factors putting the Highlands out of ecological equilibrium."

Available evidence suggests that foxes eat more lamb on poor sheep ground and in years of bad lambing than on good sheep ground in years of good lambing. The same thing can be said of the Golden Eagle. During a fifty-one-day period I found that on a good sheep hill neither the eagles nor the foxes had any lamb in at all. The eagles in fact were killing fox cubs.

The fox is a considerable carrion eater and it is the presence of carrion that makes possible the density of foxes in certain parts of the West, where natural prey is scarce. A hundred years ago Osgood Mackenzie was killing bags of hares, grouse and ptarmigan on the Western hills—where hardly one of these species is to be seen today. Yet there are plenty of foxes on the ground—one to every eight thousand acres on a recent census. What do these foxes live on? It has been shown that the average yield of lambs is sixty five per hundred in good years and thirty per hundred in bad years, with many ewes dying in addition. This provides a bonanza for a scavenger like the fox.

Everybody knows that some foxes kill lambs and that some foxes don't. Everybody knows that lambs are killed by foxes in some places and not in others; but nobody knows how many lambs foxes kill and in what circumstances. Very little account is taken of what else they do. I propose to look at some of the other things Highland foxes do.

No animal has shown a more adaptable capacity for survival in the face of constant human pressure. All the killing—trapping, shooting, poisoning, snaring, dogging, terriering, gassing—has failed to get to the back of the fox. He has his ups and downs, his cyclical fluctuations, whatever human beings do or don't do. In some years litters are big; in some they are small; but nobody yet knows why. Man prunes the fox stock, as he did with the rabbit before myxomatosis, and the fox thrives. Old Reynard Cat's Eyes—Tod, Charlie, Charles James; call him what you

Highland Dog Fox
Dog foxes and vixens look alike and it is difficult to sex them in the field unless one has a practised eye and the animal is close at hand, or one is looking at it through binoculars. A good way to practise your ability in this respect is to look at a sheep dog at work on the hill and try to tell its sex as it moves about. Dog foxes mate as yearlings and vixens breed as yearlings. Although it is impossible to lay down hard and fast rules, the evidence suggests that young vixens whelp later than older ones. This is certainly true where red foxes are experimentally bred.

will—will be here for a long time yet. The MacQueen who killed the last wolf with a dirk is going to be a lot longer dead before some future MacQueen whuttles the craig of the last Highland fox.

Once upon a time a favourite conundrum was: what is the connection between cats and clover? The answer was that the cats killed the field mice, that killed the bumble-bee grubs, that would have become bumble-bees, that would have fertilised the clover. So where you've lots of cats you've lots of clover. Nowadays I find myself asking myself about the relation between foxes and the Clearances, the tentative answer being that sheep replaced people and the old Highland cattle husbandry, and now we have a lot of lambs with foxes preying on them, whereas in the old days, calves were outside the fox's prey range. The fox in the Highlands is a problem of modern land-use.

Highland foxes kill rats, rabbits, hares, field mice and field voles. They kill frogs, earthworms, beetles, hares, capercaillie, plover, blackcock, weasel and stoat. They like carrion—dead sheep, dead deer, dead anything. They eat blaeberries, raspberries and brambles.

Foxes in an area populated by field voles will eat mainly voles—up to a dozen and more in a day. We know that a vole weighing an ounce or so will eat two ounces of grass a day and Jim Lockie has shown that voles in Wester Ross eat twenty-three pounds each in the winter months. Foxes probably eat two thousand voles a year.

The fox bolts voles like a man bolting porridge, so there is hardly ever any evidence of vole remains at a den unless the vixen brings in more than the cubs can clear at once, which is not very often. Evidence on predation comes from the fox's droppings, which contain the indigestible residues. This can show vole as high as eighty per cent in years of high vole numbers. Lockie, working on sheep ground in Wester Ross, found that droppings contained thirty-one per cent vole by weight in summer, and twenty-six per cent in winter.

Nowadays we have some idea of the amount of carrion eaten by foxes. It isn't a great deal of information, but it is hard information and much better than the intuition. In Wester Ross, Lockie found that foxes ate sixty-three per cent carrion in the winter (twenty-five per cent adult sheep and thirty-eight per cent adult deer) and in the summer five per cent adult sheep carrion. In the same place, during the summer, fifteen per cent of the fox's food was lamb, taken dead or alive. In certain parts of the West it is carrion that makes possible the density of foxes.

The sexes in foxes look alike and it is almost impossible to tell one from the other in the field. Dogs are bigger and heavier than vixens. Once you have been looking at a den for some time you are soon able to tell one from the other. The average weight for dog foxes is usually given as around fifteen pounds and for vixens twelve pounds; but many Northern foxes go far beyond these weights. Size varies widely and some Mountain foxes look as big as sheepdogs. In the autumn, cubs of the year can usually be distinguished from adults by their lank, long-legged appearance.

Breeding begins early in the new year and this is when foxes are most vocal. February is the commonest mating month and cubs are born fifty-one or fifty-two days later. There is only one litter a year, usually from four to six cubs. I once found thirteen cubs in one den but it turned out that two vixens were sharing it. Both were killed by terriers.

The vixen chooses the den and scrapes it out before she is due to give birth to her cubs, which are born on the bare earth at the end of the burrow. Occasionally both adults can be seen digging. The vixen will clean out several sites before fixing on one. Many Highland dens are traditional, and the fox-hunter goes round them each year. In mountain areas and sea cliffs, rock holes and cairns are much used. In the Scottish Highlands these rock strongholds can be fearsome places, with boulders as big as houses, great slabs of rock piled one on top of the other; and cracks or holes thirty or forty feet deep. Some of these cairns can extend for half a mile, providing the foxes with a veritable fortress in which any terrier can become lost. Sometimes Highland foxes will den up in deep heather or bracken, and a vixen sometimes has her cubs there. It is often difficult to tell whether such sites were her first choice, or whether she moved the cubs after a disturbance somewhere else.

A vixen will readily move her cubs if the den is disturbed. Sometimes she will move them for no obvious reason, or perhaps because the old site has become foul. Cubs are frequently moved at weaning time to grow on somewhere else to the age of independence. Highland vixens often yard up with their cubs in heather far from the old den, or any den, and here the family can sometimes be seen by day or put on foot. Such moves, made when the cubs are able to follow on foot, are very much of a ritual with some mountain vixens.

A big vixen can be a terrible handful for the most varminty or hard-bitten terrier in a teeth to teeth chopping match. I have seen terriers scarred and bitten through the face before the vixen would bolt. Big cubs will grapple with, and chop, the terrier that is worrying them. The vixen is shot as she bolts; then the terrier kills the cubs. Somebody waits up for the dog fox and the whole family is wiped out.

Dog foxes, like dog wolves, are good fathers and carry prey to the vixen when the cubs are small. If the vixen is killed the dog may rear the cubs himself—if they are not too small to eat flesh. Cubs play as soon as they can walk and right up to the time of dispersal. A grouse wing or a hare's leg is a favourite play-thing but I have seen other things used—a dead mole, an old boot and a hen's leg.

Outside the breeding season the fox is usually solitary but there are exceptions as when three or more gather at a carcass during a hard winter. I have seen seven together, and three at the body of a stag. I once saw four stalking after a sickly yearling roe buck but this was in the breeding season when more than one dog fox will follow a vixen. The idea that foxes hunt in packs has no foundation in fact.

During the rut foxes become more vocal: then the triple bark of the dog can be heard nightly. The vixen has a squalling cry which is sometimes uttered by the dog fox. Cubs at play make a variety of puppy noises.

Dog foxes and vixens have scent glands under their tails. They deposit their scent all over their range—on logs, tufts of grass, heaps of stones, hummocks and mole heaps, as well as with their droppings and urine. The tailed droppings of the fox are usually found on such prominent places and are easily identified. I once followed a fox on a mile circuit in tracking snow, and found scent and urine markers at regular intervals.

Normally a fox walks or trots; but he will run fast and far when he has to. Many foxes are caught in traps and some are left to die and rot in them. Fortunately the gin trap has been illegal since 1st April, 1973. I have found a vixen caught by a foot in one trap and by her brush in another. In another I found a foot which the fox had chewed off to make its escape. Such a fox would probably be difficult to trap again.

Some foxes have more trap sense than others and are more difficult to poison than others. Although always alert and suspicious, with an excellent nose and ears, and good eyes, a fox with cubs can be watched at close range by a watcher if he is off the ground. I have watched a vixen without trouble under such conditions.

The fox is found throughout the Highland mainland, but is absent from all the islands except Skye.

Opposite

Unlike the brown hare, the Mountain Hare uses burrows and other shelters. It is the only Scottish mammal, apart from the stoat, that turns white in winter.

Left

Adult Brown Hare in young coniferous forest. Although brown hares are usually animals of open country, they are often found in woodland and many leverets are born there.

Right

Vixen hunting voles among heather. It has been shown that Highland foxes probably kill about 2,000 voles a year. Man wages constant war against foxes in the Highlands without, in any way, affecting their numbers.

Below

Soon after birth, the leverets separate and each occupies a form of its own. The doe visits them in the night and calls them to her to be nursed. If she has to carry them, she does so like a cat carrying kittens.

Left

Fox cub at the age of three weeks on its first sortie from the den.

Right

Common Shrew eating an Earth Worm. The shrew immobilises the worm by biting it along its length. It can then eat its fill and return to the worm in an hour or so to feed again.

Below

A Vixen with well-grown cubs loses much of her bloom. Her coat becomes touzy. A vixen may carry food right into the den to her cubs or bury it and call them out to look for it. Whether she buries it to force them to look for it is another matter, because it is a habit with foxes to bury what they can't eat at the moment.

Left

The Pygmy Shrew is the smallest shrew and the smallest Highland mammal. It is not easy to distinguish from the common shrew at a glance.

Right

The Water Shrew is the biggest British shrew and is usually black on the upper side and white on the under side. It likes clean water and is found in burns, rivers and pools where it preys mostly on aquatic invertebrates. It will sometimes be found in woodland some distance from water.

The Mole is wonderfully adapted for his underground life, having a barrel-shaped body, a sharp snout, spade-like forefeet and fur that cannot be rubbed the wrong way. The mole is a problem animal. Arable farmers dislike it because of the heaps it puts up; gardeners dislike it for the same reason; and it is not welcomed on golf courses or bowling greens. Mole runs are often used by weasels changing ground.

Mole

Mole with Earthworm
Although the mole catches most of its prey underground, it will surface to take worms on a night when worms are surfacing in numbers. Badgers hunt earthworms at the same time in the same places, so the mole sometimes becomes the prey of the badger.

Everybody knows the mole, if not by sight, at least by its handiwork—the great scattering of mole-heaps in fields and hill grasslands. The mole throws up these heaps when he is driving his communications tunnels and digging for food. He hunts while he tunnels then uses the tunnels for moving about in. The speed at which he tunnels depends on the nature of the soil and the amount of prey he is finding. In a rich seam, he eats often and drives on slowly. In poor soil, he digs much for less return. A great number of mole-heaps does not indicate a large number of moles; merely a great amount of work.

The mole needs a good soil, rich in earthworms and other food. Most moles are found on good arable land and in deciduous woodland. But in the Highlands, they will be found up to 1,500 feet on good grassland or on re-seeded areas. They avoid peaty moorland until it is broken up and re-seeded; then they move in. They are not found in coniferous forest.

Because the mole spends most of its life underground, few people ever see one alive and free. But the animals do surface periodically. A mole driving a tunnel among grass roots will do so from time to time. A long spell of drought will bring one to the surface, sometimes in good daylight. Moles are more readily seen above ground in early spring and in late summer than at any other time. Signs of them can be seen when a mole-heap moves; the mole underneath is pushing up soil. Or sometimes you will see movement when the mole is burrowing just under the surface, causing the turf to heave and fracture.

Mole tunnels are not all driven at the same depth. Most of them are less than a foot below the surface but some go much deeper. The mole fits them tightly and constant use packs the walls hard, so that they remain intact for a long time, even when they are no longer being used. Voles and shrews use the runs at times, and a weasel will travel through them to get from one part of a field to another.

Among the mole's ordinary working heaps you will often see a much bigger one—two feet high and several feet across. This is the so-called fortress heap. In the old days, many drawings were made of its interior to illustrate this. In actual fact there is no particular plan, although the heap contains tunnels and bolt-holes.

In such heaps the mole makes its sleeping nest and, occasionally a female will have her young there. But the breeding nest is usually below ground, with nothing overhead to betray its position. The nesting chamber is an enlargement of a tunnel from which the mole pushes the excavated soil to the nearest shaft. Such nests lie deep under the unbroken surface.

Gillian Godfrey developed the technique of finding moles by tying a radio-active ring to the animals' tails and keeping track of their movements with a geiger counter.

The mole feeds heavily round-the-clock and round-the-year, and dies quickly if deprived of food for more than a few hours. He hunts when he is hungry, and sleeps between hunting forays. Permanent grassland and woodland are the most durable habitats. On arable ground the moles come and go. Each animal has a range that it defends against other moles, the size of the range varying according to the kind of habitat, the age and sex of the mole, and the time of the year. Ranges are smallest on the best ground.

The breeding season is from January to May and is usually over by June. There is only one litter a year. At birth, the young moles are naked and red-skinned and have the big fore-feet and heavy shoulders of the adult. Their fur begins to grow in a fortnight and a few days later the nestlings are fully clad in a silvery-black coat. Their growth rate is related to the amount of food available.

The mole is widely distributed in the Highlands up to 3,000 feet. It is found on Skye and Mull, and has been introduced to Ulva.

Common Shrew

Shrews are bundles of nervous energy. They live at high pressure and age quickly and most of them die before their second autumn. They are active and restless, ever on the hunt, rummaging, scraping, burrowing and probing in the ground litter, requiring great quantities of food and sleeping between bouts of eating. They die if deprived of food for more than a few hours.

The shrew's strength is out of all proportion to its size, and an animal weighing less than half an ounce can quickly drag into cover a vole three times its weight. Active after a short rest, the shrew will move a mole-heap of litter in a few minutes of commanding gluttony. He rustles as he rummages without thought of stealth, and he squeaks as he rustles. Outside the breeding season, he leads a solitary life, but he is always vocal, always quarrelsome, and frequently aggressive towards any other shrew he meets, whatever the species.

When shrews do meet, they sometimes have brief contact, followed by retreat. At other times, there is argument and quite often there is a battle; but such battles are rarely fatal. A fight to the death, when it does take place, would seem to be a fight between rival males competing for a female. Other fights, whether savage, or mere grapple and bluster, are very likely territorial, and the beaten animal is allowed to retire on its feet. There is no doubt at all that shrews behave aggressively at the mere sight of each other and if they were as big as wildcats, they would be terrifying animals indeed.

It isn't difficult, if you are in the right place at the right time, to hear a shrew squeaking as it goes its rounds, at any hour of the day or night. It is particularly easy in spring and early summer because of the heightened activity of the breeding season. Two or more shrews together mean more excitement and quarrelling and, therefore, more noise. Where you come across a small group, foraging amicably together, and uttering only squeaks of pleasure, they will almost certainly be a female with weaned young still on her territory.

Like so many other animals, the shrew uses regular runways on his territory and knows his way about without having to look where he is going. He doesn't have to think about where the next turning is. He can find every one without effort, just as a man can find a familiar door knob in the dark. This is the kinaesthetic sense, enabling an animal to leap without looking or find its route without thinking.

The shrew can go where he is apparently too big to go. He uses the burrows or tunnels of voles and mice, but also makes tunnels of his own, and he will drive surface creeps after rain. His sense of smell isn't good but his big whiskers, sensitive to contact, help him to steer his course in remembered runways. He finds his prey almost by falling over it, not smelling it until he has made contact with his whiskers. But he will recognise a scent if it is strong enough and his nose is close enough to it. Thus the blood scent of an earthworm drawn across his path will pull him up short. Fresh blood smell of any kind will halt him in his tracks and start him seeking. Blood smeared on the stem of a bush will get him off the ground.

It is almost certain that shrews store surplus food, as moles do. The stored prey is disabled to immobilise it. In the case of earthworms, the shrew pricks them with his teeth all along their length so that they cannot move from the store, but remain alive and fresh to be eaten later. Where you hear shrews squeaking, you should be able, with patience, to see them by parting the ground cover, although at first the only sign may be a heaving of the litter. Sooner or later, the shrew will break out, questing with its flexible snout, pouncing on a woodlouse here, a spider there, or an earthworm uncovered by the upheaval.

The shrew's breeding season is from March to October. By October, most of the old animals have died off. The female makes her breeding nest with grasses and whatever else is lying to hand. It is closely woven and substantial, and may be made under a tree root, a hedge-bottom or in a hole in a banking. From the nest, a bolt-hole leads into the surrounding litter or the underground tunnel system.

The young shrews grow quickly, but development is slow. Their eyes don't open until they are close on three weeks old. They are weaned a few days later. The young don't usually mature until the following spring.

Many mammals kill shrews, but not so many eat them. Dogs, cats, foxes, martens, stoats and weasels kill them. Dogs and cats that swallow them usually vomit them soon afterwards. Foxes, stoats and weasels can digest them, but appear only to eat them when the alternatives are shrew or nothing. Jim Lockie has noticed that the pine-martens of Wester Ross, like stoats and weasels in plantations, rarely eat shrews even when they are there in abundance. This has been my own experience. Bird predators, on the other hand—tawny owls, barn owls, short-eared owls, kestrels, buzzards—eat shrews readily and carry them to the nest when feeding chicks.

The distastefulness of the shrew, associated with the musky odour from its flank glands, doesn't prevent it from being killed, even when the predator has no intention of eating it. Why then should any predator kill a shrew at all? One has to assume that shrews are often killed in error or for sport. There is no evidence that sickness following a meal of shrews puts any predator off killing other shrews.

The common shrew is abundant throughout the Highlands, but it is not found on Orkney or Shetland, in the Outer Hebrides and some of the Inner Hebrides—notably Eigg, Muck and Rum. It is found on Islay, Jura, Gigha, Mull, Ulva, Skye, Colonsay, Raasay, South Rona and Scalpay.

Pygmy Shrew

The Pygmy Shrew and the Pipistrelle Bat tie for the title of Britain's smallest mammal. The weight of pygmy shrews varies from a tenth to a fifth of an ounce. But this isn't much help if you are trying to tell the pygmy from the common shrew which looks the same and isn't very much bigger. The pygmy's tail is about two-thirds of its body length; the common shrew's is half or less. Both species have hairy tails, but the pygmy's is thicker.

Pygmy shrews are found from sea level to the tops of the highest mountains, wherever there is plenty of ground cover and enough invertebrate food. They are hardly ever as numerous as the common shrew on the same ground, and don't become very numerous even where there are few, or no, common shrews.

The pygmy occupies the same kind of ground as the common shrew and forages on and in the ground litter, squeaking as it prowls and probing nervously with its flexible, trunk-like snout. It is an altogether more nimble and active animal, a volatile atom, with the snake-dart and recoil of a weasel. It can out-run, out-climb and out-jump any common shrew it happens to meet; but, before running away, it will squeak back defiance at the threat of assault.

Although it is an insect eater, it is sometimes frightened off by insects; for example, a very big beetle threatening counter-attack. Very big earthworms can defeat it; then it is like a man trying to hold an anaconda. Depending on the nature of its prey, the pygmy will eat up to twice its own weight in food every day.

Like the common shrew, the pygmy leads a round-the-year and round-the-clock existence, and animals born in the spring and summer of one year are all dead by the autumn of the next. Its life is mainly eating and sleeping. Unlike the common shrew, the pygmy doesn't burrow and, when it wants to get below ground, it uses the tunnels of mice and voles. It makes its nest of dry grass and moss, under a stone or in a dry-stane dyke, in the cavity of a tree stump, or even in a grass tussock.

The breeding season is from April to

Pygmy Shrew
The Pygmy Shrew is the smallest animal on four legs in Britain but it manages to get just about everywhere, and has been found on the top of Ben Nevis. Like other shrews, the pygmy has musk glands that give it an unpleasant odour. Mammalian predators kill it but don't eat it; predatory birds kill it and eat it. Like other shrews, the pygmy has been the subject of many superstitions, notably that its bite is poisonous and that it can kill men and cattle. In fact, although the shrew is able to kill animals bigger than itself, it is useful and harmless to man.

August—perhaps a little later. Two litters a year are probably the rule, and the average number of young is six.

Owls, kestrels and buzzards kill pygmy shrews and eat them. Many predatory mammals also kill them, but usually reject them.

Although it is not the most numerous shrew in the Highlands, the pygmy is the most widely distributed, being found throughout the mainland and most of the islands. It is absent from Shetland, North Rona and St. Kilda.

Water Shrew

Burnside and riverside are the habitats of the water shrew. Before so many burns and rivers became polluted it was a much commoner animal than it is today. Happily this is not so true of the Highlands generally, where most of the water is still clean. Where dirty water is, this shrew is not. But you will sometimes find it, even in areas where the water is polluted, living in woodland or moorland, far from the nearest stream. There it haunts ditches and other boggy areas, breeding successfully, and feeding on the same kind of soil invertebrates as the common shrew.

When it is living near water it preys upon water and land insects, crustaceans, snails, small fishes and frogs. Like the dipper, it can hunt the burn bottom, wiggling along like a tadpole, while turning over pebbles with its forefeet in its search for caddis grubs and other prey. In quiet backwaters it hunts pond skaters and whirligig beetles, out-swimming them or rising under them in the same way as the pike comes up to take a shrew. The charge of eating fish ova is laid against the water shrew. And it does take them. But the effect of such predation can be nothing other than insignificant.

The shrew in turn is preyed upon by foxes, cats, owls, herons and stoats. Owls eat shrew, but mammals usually reject it, unless the alternatives are shrew or hunger. The pike can be a considerable predator, and certainly eats shrew.

Habitat alone is not what gives this shrew its name. It is specially adapted for swimming, and hunting in the water. Its haired tail has the sweep of an oar, and the bristles on its feet act like the otter's webs. The shrew swims high in the water, with lateral movements of tail and body, and paddles with alternate strokes of its feet.

It is a good diver, and a strong swimmer. Although it can enter the water as quietly as an otter, it is more likely to dive with a distinct *plop*! When it is swimming underwater it often looks silvery because of the air bubbles trapped in its fur. It isn't difficult, if you sit quietly and for long enough, to watch a shrew working underwater, swimming close to the bottom, probing for prey with its flexible, trunk-like snout, and kicking up clouds of ooze when it pounces.

The water shrew is usually, but not always, a black and white shrew: black above and white on the underside, with a clean line of demarcation along the flank. But in some animals the fur is more brown than black. The tail is dark brown on the upper side, whitish on the under, where there is a fin of swimming hairs. A white eye patch is common, but not all water shrews have it.

The nest of the water shrew is usually in a bank not far from water, but may be in woodland some distance from it. Nesting material is usually grass and withered leaves, and the nest has upper and lower exits in the bank. The breeding month is May, and young are born in June. Two litters are probably reared, and the season extends to September.

All the shrews are strong fot their size, and this is the biggest and most powerful of them all, being able to kill small fishes, small frogs, and very big snails. Like the others it hunts round the clock, and round the year.

The water shrew is found in the Highland mainland, but is absent from all the islands except Skye, Kerrera, Islay and Arran.

Long-eared Bat in flight.

The Field Vole is notable for its four-year population cycle. Every four years it builds up to a population peak; then there is a crash. This cycle takes place only in simple habitats—for example in coniferous forests. Elsewhere, in complex habitats, with great variety of plants and wildlife, there is no cycle of abundance and scarcity, and the vole's presence is hardly noticed.

The Bank Vole is more mouse-like than the field vole. Its ears show through its fur and it has a longer two-coloured tail—dark on the upper side, pale on the underside. As its name implies, it is found on banks and slopes with plenty of ground cover.

The Stoat in Winter dress is known as an ermine. Whether the stoat changes completely or partially, the tip of the tail always remains black.

The Water Vole
Although frequently mistaken for the brown rat, the water vole has a blunter face. It is more robust and more heavily built. Its ears are barely visible, and have a lid for keeping out water. The sexes are alike. The water vole is an accomplished diver and swimmer but, oddly enough, it is not as fast a swimmer as the brown rat. When not afraid, it swims with the whole of its head and its back showing above the water; when it is afraid, only the top of its head shows. It can perform dives lasting about twenty seconds. Although sluggish waterways are the common habitat of the water vole, it will be found on clear, fast hill streams in the Highlands. In very severe weather, it will lie up for several days at a time as other voles are known to do. It stores food in the Autumn. It usually surfaces under cover and will even bring up cover with it in the form of dead leaves, and these screen it as it breaks the surface. Young water voles are able to swim and dive as soon as they leave the nest.

Left

Nest of young Weasels with their eyes just beginning to open. This nest was in an old dry-stane dyke. The bitch weasel removed her family soon after the nest was uncovered.

Below

Pine Marten emerging from a burrow in the snow.

Pipistrelle Bat

Pipistrelle is the bat of many names. The Scots word is bauckie bird; flittermouse and airymouse are English equivalents. It is the commonest bat and the best known, which means no more than that it is the one most people see.

Besides being the commonest British bat, pipistrelle is the smallest, running neck and neck with the pygmy shrew for the title of Britain's smallest mammal. Its wing span is a mere hand's length and it weighs under a quarter of an ounce.

British bats fall into two groups—those with a nose-leaf and no tragus, and those with a tragus and no nose-leaf. The nose-leaf is a flap of skin that grows from the upper lip and covers part of the face. The tragus is a lobe of skin attached to the base of the ear-trumpet, and looks like a small inner ear. Pipistrelle has a tragus but no nose-leaf.

In midsummer, bat-light for pipistrelle is sunset when it will be seen, hawking around, before the owls are on the wing. An overcast night will bring it out earlier but a bright night doesn't keep it in. It is not unusual to see it on the wing in the month of June, in very strong light.

The flying season is from March to October, but pipistrelle can be seen in any month of the year in sunny weather, even when the hills are blanketed with snow and the drifts are six feet deep. I watched one on the second day of March when the snow-drifts were crusted in the frost. It was flying along the middle of the road, weaving from side to side, and wavering among the trees, then it breasted a snow-drift, causing a puff of white, before disappearing into a ruined house.

The hunting pipistrelle will hedge-hop like a sparrowhawk—wavering, criss-crossing, jerking up and down; or it will flutter up to tree height and fly at speed erratically at that pitch. It is a fast and capable flier despite the snatches and breaks in its flight and its sudden changes of direction. It will come down to a pool to drink, pitching on the edge to sip. At other times, it will sip water on the wing, hovering like a kestrel just above the surface. A bird-bath in a garden will attract pipistrelle either to drink or splash in.

On the wing, pipistrelle is talkative, although the human ear eventually gets past the stage of hearing. Where many bats are flying at the same time, they can be as noisy as swallows.

The other sound uttered by pipistrelle is inaudible to the human ear. This is the short-wave squeak that the bat uses for echolocation. It finds its prey by uttering these squeaks, which are then reflected back from any object in its path. Thus pipistrelle, like other bats, avoids collisions in the dark, and can weave its way through branches and round obstacles in its path.

The mating season is in Autumn and Winter. The female pipistrelle stores the male's sperms and when she produces her egg the following May, it is fertilised from this store. A single young is born in the second half of June or early in July.

Probably pipistrelle takes all its prey on the wing, and the main prey appears to be gnats. Gnats, and insects of this size, are eaten on the wing; but larger ones are often taken to a plucking place and dealt with there in the same way as a sparrowhawk uses its so-called plucking stool. Sometimes the bat will put big prey into its pouch, which is situated in the interfemoral membrane—the flap of skin that stretches from legs to tail.

People do not usually bother much about pipistrelle, but some will clean out a roost because of the dung that accumulates below it. Large numbers of pipistrelles will often roost together.

Apart from man, pipistrelle has few enemies. The tawny owl is sometimes a considerable predator.

Pipistrelle or Common Bat
This is the commonest species, found practically everywhere. The sexes are alike. It is the smallest of British bats. It is a very active species and, in warm winters, can be seen out and about almost any day. Full hibernation lasts from mid-October to mid-March.

Daubenton's Bat

Daubenton's Bat. Illustrated below
In this species the sexes are alike. Although it is gregarious, it can sometimes be seen singly or mixing with other species of bats. It flies about an hour before sunset, and probably hunts all night. Daubenton's is found in almost all Scottish counties as far north as Elgin.

This is a bigger bat than pipistrelle, with about two inches more of a wing-span; but this isn't an easy thing to notice in the field. For all practical purposes, the bats are about the same size. There is another way in which Daubenton's can be confused with pipistrelle. Pipistrelle hunts over water some of the time, while Daubenton's hunts over water almost all of the time, so they are often in the same place together. Daubenton's is the true water-bat, the one usually taken on the fisherman's fly.

The food of Daubenton's bat is almost entirely aquatic insects caught in flight and eaten on the wing. Larger insects, like dragonflies and big moths, are pouched and taken to an eating place.

On warm, humid summer nights, large numbers of Daubenton's can be seen hunting together, hawking low over the water like swallows. Every now and again, one of them will lightly break the surface as it drinks on the wing. A ground mist will put the bats higher, while a cold wind, or a sharp drop in temperature, threatening frost, will send them to roost—perhaps for the rest of the night. In fine weather, they will hunt on and off through most of the night, going to roost shortly before sunrise. They are not put off by light rain. This bat roosts in holes in trees or rocks as well as in caves and buildings. Bats roosting in a hole in a tree will squeak and sizzle if the trunk is slapped about the time they are due to come out. Daubenton's is not an erratic flier. It hunts in circles, with wings flickering like a curlew's when the bird trails from her eggs. As it always hunts over water, it can be confused only with pipistrelle. It doesn't like daylight, but will fly by day if disturbed at its roost.

The Daubenton's female gives birth to a single young in June or July. The bats hibernate in caves or buildings. They don't hibernate in trees, although they roost there in summer.

Long-eared Bat

This is quite a common bat, easily recognised, even in flight, because of its enormous ears. When at rest, it folds its ears along its flanks, under its wings, leaving the tragus erect. When half awake, or alerted, it partly unfolds them so that they curve up and over like a ram's horns. This bat has good hearing, and almost certainly uses its ears when hunting.

Bat-light for this species is after sunset—from half an hour to an hour later, depending on weather and time of year—and it hunts on and off during the night. On good flying nights, there will be bats on the wing throughout the darkness, squeaking as they hunt. The flight is moth-like, with occasional glides, and the bat will hunt from hedge-height to tree-height, hovering among the foliage like a bee at a foxglove bell, with ears directed well forward, listening. It is also agile on a roof or wall. My friend, Jim Lockie, watched one running about like a spider, catching moths on a byre roof that had been stripped of tiles that day.

When hunting, the long-eared bat alights frequently to dismember prey, and the strippings from beetles and other insects can be seen at such places. It appears to be a very thirsty bat because it alights at regular intervals to drink. It flies to roost about an hour before sunrise.

In summer and early autumn, long-eared bats often hang out in trees. They move to more sheltered quarters in houses or hollow trees as the weather grows colder. Summer roosting places seem to attract a succession of long-eared bats, just as certain holes in trees attract generations of tits, or certain rocks attract otters. As soon as you remove one animal, another takes its place.

The hibernation period is from October to March. The bats sleep singly or in small groups, but stir readily and irritably, and often fly on mild winter days.

In July and August they assemble in large numbers, and there is considerable evidence of migratory movements. Mating season is October/November, and April/May. The first mating is of adult animals; the second of animals too young to mate the previous autumn. Long-eared bats have one young in a year.

Water Vole

The water vole sits on a raft of rushes under the overhanging bank, his fur shark-toothed by the shadows of grass blades drooping from the top. He is feeding one-handed, nibbling on a grass stem held in one fore-foot and stroking his flank fur with the other. Many short, bitten stems lie on the raft beside him. Presently he drops the one he is nibbling and cranes forward, listening. The farm terrier is out and coming along the bank.

Plonk! the water vole dives from his raft; uncorking the bottle as we used to say. He scrabbles on the bottom, stirring up a concealing cloud of mud, grit and leaf flinders. Then we see him, sleeked out like an otter, heading into the underwater entrance to his burrow.

Fifteen minutes later, the blunt face, with boot-button eyes, appears alongside the raft again; but this time the vole doesn't mount it. He launches away without a splash and swims towards me, coming ashore beside a great clump of rushes, when he is lost to view. Across the stream, on the bank top, the grass begins to move and I have a brief view of another blunt-face, followed by a chestnut body. This vole is feeding beside a bolt-hole, and will be the big vole's mate.

Before long, the big vole under the bank reappears in the water and swims back to the opposite bank, trailing a wake and a length of greenstuff. He disappears into a bank burrow with the green length in his mouth.

This is water vole territory and the pair occupy a stretch of the stream a hundred yards up and down. Along this length are many burrows, but no other voles; the territory holder defends his ground fiercely, allowing only his mate and his offspring within it. Upstream and down he has marked his territory with scent from his flank glands, and little heaps of droppings are strategically placed where no intruding water vole can possibly miss them.

Yet the big vole will lose his territory in the autumn, and his life before the winter. He will lose the first to one of his growing family, and the second because he will be senile. Although he is so much bigger than the field vole, his life span is about the same, and few of his kind survive a second winter.

The water vole is frequently mistaken for the brown rat. Summer rats invade water vole burrows, taking them over and killing the occupants; so the two species are often seen in the same place at the same time. Hence the basis for confusion.

One day when I was out with my terrier, I saw a brown rat leaving a water vole burrow, then scurrying along the water line, following the well-trodden runway of the voles. The terrier ran the rat's line without check, dug her from her bank nest, and killed her and her seven young. No doubt the rat had been doing execution among the young water voles. It is ironic that the water vole should be so frequently mistaken for, and named after, one of its greatest enemies, and persecuted by man as a result.

I have many times watched a weasel popping in and out of water vole bolt-holes on top of a banking. The weasel is a persistent vole hunter and probably kills a lot of young in the nest before the mother can carry them away. I once saw a stoat coming out of a water vole burrow with a young one in its mouth.

Bird predators are the tawny owl and the heron. The heron stands still like a carving until a vole appears, then suddenly his snake-neck is unleashed and his dagger-beak stabs the water. The bird shakes his head and stalks ashore with a vole securely pincered. The vole is dabbed against a stone and gulped down, then the heron flaps away to try another stretch of water.

Mortality is high among young voles, and many die before they have seen much of the outside world. But enough survive through the winter to become the territory holders and breeding stock of the following spring.

In the autumn, they lay up their store of

green food in special chambers of the burrow system and draw on these during prolonged spells of frost or blanketing snow.

Although slow-moving rivers, ponds and canals are typical water vole habitats, the animals will also be found in mountain streams up to 2,000 feet.

The breeding season starts in the second half of March and lasts into late September or October. There are usually four or five young, and there may be four or more litters in the season. The summer litters are the biggest. First young appear in May, and the young from these litters will soon breed themselves.

Waterside grasses and seeds are the main food of the water vole but it eats many other vegetable foods, including wild garlic, iris roots, beech mast and stubble grain. Water snails and fresh-water mussels are occasionally eaten. Good baits for attracting water voles into the open are apples and carrots.

If a man sits quietly in an area obviously occupied by water voles, he should soon see them. The young, especially, play much in the water, and are not difficult to watch.

The water vole is widely distributed on the Highland mainland. In the Highlands, the race of water voles is smaller and darker than elsewhere and has been given sub-specific status.

Weasel
The Weasel is a specialist predator on the field vole and, in young forest plantations, lives on this prey almost exclusively. In such a habitat the weasel, like the vole, makes a good subject for a field study. This weasel has been marked on the ear for identification purposes. The technique of marking is to box-trap the weasel, anaesthetise it, then ear punch it, allowing it to make a full recovery from the anaesthetic before releasing it. A bitch weasel will build the skins of field voles into the nest where she has her kits.

Short-tailed or Field Vole

The eagle sweeps round the corner of the wood with one wing down and pitches on a boulder in a young spruce plantation. He looks right and left, hops to the ground, thigh deep in grass, and begins to walk, high-stepping and flat-footed as the eagle always is on the ground. Presently he starts dabbing this way and that, and gulping. He is catching voles, small prey for so large a bird; but the area is swarming with them and they are easily caught.

A quarter of a mile away a small falcon is hovering above the slope as though suspended from an invisible wire. The bird is a kestrel hunting, like the eagle, for voles. At a height of fifty feet he can spy the slightest movement on the ground below.

The kestrel pays no attention to the eagle. They are neighbours with different interests and don't interfere in each other's affairs. Not for the kestrel the high reconnaissance flight and the mighty swoop. He works his hunting ground section by section, hovering for a minute here and a minute there, with wings fluttering, tail fanned downwards, and eyes fixed on the ground below.

Presently the kestrel sees movement in the grass and his fluttering stops. He tilts and dives with wings half closed and, when it seems that he must crash into the ground, he opens them again and stretches down, clutching with a slim, yellow-taloned foot. A fat, furry vole squeaks and dies without ever knowing what has struck it. And the kestrel stands over his prey with wings down and glares about him before carrying it away, clutched up against the feathers of his breast.

Another kestrel is perched on a roadside post—a tapering silhouette, round-headed, round-shouldered, and long-winged. She is not sleeping or resting. She is watching the tussocks on the forest edge. Her big round eyes, dark and lustrous, have the sun-glint in them. In the matted tussocks voles are swarming, which is why she is waiting on. She has probably been eating voles, and nothing but voles, for weeks on end.

Nearer the road a short-eared owl is quartering the ground, drifting and slow-flapping against the soft back lighting of the hazed sun. The owl upends suddenly and swoops into the grass, where he stands with wings half open. Then he beats up again, wide-winged, and flaps to an old post with a field vole clutched in a foot.

On the far slope another hunter is stalking voles—this time a mountain fox, big-boned, with thick ruff and white-tipped brush. He is prowling among the tussocks as stealthily as a cat, sniffing here and listening there, and occasionally exploring likely places with a black paw. He looks up when the eagle flies over, then returns to his stalking. Suddenly he leaps up and pounces with fore-paws together. Then he lifts one paw and reaches down to snap up the vole held under the other. He chops it, swallows it at a gulp, and begins to stalk again.

All over the forest the rough grass is honey-combed with the entrances to vole creeps. A weasel pokes out from one of them, peers at us curiously for a moment, then darts in to get on with his hunting. Vole trackways are everywhere, leading to creeps and forming mazes like railway junctions. We see vole after vole scurrying along the tracks—a sure sign that their numbers are high. We look under tufts of grass for their little platforms of chopped grass. These platforms are sometimes used as latrines; more often the droppings are scattered along the runways.

The field vole is a grass eater, and young plantations make ideal habitats for them because of the abundance of grass as food and cover. In such situations their numbers fluctuate on a four-year cycle, resembling the cycles of the lemmings of the Arctic. The number of voles on an acre of such ground can be as few as five or ten or as many as six hundred or more. The ups and downs of voles are spectacular enough, but the most striking thing about them is the regularity of their

Field Vole
The field vole might be described as the bread and butter of the hunters. It is preyed upon by many species of mammals and birds. Mammal predators are fox, badger, wildcat, stoat, weasel, polecat, mink, otter and pine-marten. Bird predators are golden eagles, buzzards, kestrels, hen-harriers, tawny owls, long-eared owls, short eared owls, barn owls, ravens, crows and magpies. The heron will stalk and kill field voles, especially where ground has been suddenly flooded. Yet this predator force can make no impression on the number of voles when they are at the peak of their cycle.

peaks, which takes place about every four years.

Cycles like this take place in habitats classified by ecologists as simple. A simple habitat is like a monoculture in agriculture, and just as subject to attack by pests or as likely to be upset. Voles in mixed woodland or mixed scrub-land do not fluctuate in this way. Monocultures of trees and monocultures of grass provide them with the ideal habitat for such fluctuations.

Predators gather where voles abound. They are preyed upon by stoats and weasels, foxes, kestrels, owls and buzzards. When vole numbers are high, the predator force increases. When the voles crash, the predator force is reduced. Stoats, weasels and others depart. Short-eared owls leave the area. An area that has had two pairs of short-eared owls for three years can have ten or twenty pairs when the voles reach their peak.

The breeding season of voles varies from one year to another. They may breed right through a warm winter but, when their numbers are high, breeding can be poor, with no young born at all. Young voles can breed from the age of three weeks, but this depends on the state of the population cycle. Voles litter regularly, and a female can be pregnant while nursing a family. The number of young in a litter varies from one to eight but is usually between three and six.

The field vole makes her nest of dry grass, which she chops and shreds, and builds into a depression in the ground at root level, or in a grass tussock. Some voles make their nests under any sheeted material lying on the ground—corrugated iron, asbestos or boards.

Once the voles reach the peak of their cycle they are doomed. There is a population crash and most of the animals die. If we visit a young coniferous plantation at the tail-end of such a cycle, we will find that the grass has a brown and shrivelled look. Whole tussocks come away in our hands, sheared through at the base by many nibbling teeth. The ground underneath is riddled and criss-crossed with runways, and brown shapes scurry for cover as we expose them. The area is derelict. The voles can even destroy large trees by barking them. Although they are not good climbers they can manage to bark older trees up to twelve feet from the ground, thus destroying forest as surely as fire or the woodsman's axe.

But the voles begin to build up again. In the early days of the build-up, the grass is green and luxuriant. The shadowy traces of voles can still be seen at root level but they are like over-grown footpaths that have not been used for some time. There are few fresh signs of voles anywhere. A solitary owl quarters the lean ground where prey was so recently plentiful. At last we find a vole nest with tiny young and stand by while the mother, resenting the disturbance, carries them off, one at a time, in her mouth.

Voles prefer rough grass where the cover is tall and thick. Grass is their food and their protection. Young plantations of conifers make ideal habitats, and large numbers are found on sheep ground. They are found on all marginal grassland and on Highland moorland, even to the Alpine plateau of the Cairngorms. In scrub and woodland they are thinly distributed.

The field vole is found on the mainland, on North and South Uist and Benbecula. It is replaced by allied races on Orkney, the Outer Hebrides, Eigg, Muck, Islay and Gigha. It is not found on Shetland, Lewis, Barra, South Rona, Raasay, Rum or Colonsay.

Orkney Vole

This species is closely related to the Continental Vole, but is similar to the field vole of the Highland mainland except that it is larger and darker. It is found on farmlands and rough grazings, including heather, up to about the 700 feet contour. Over that height, it becomes scarce. The breeding season is from May to September, and nests may be on or below the ground. Orkney voles breed more slowly and live longer than mainland voles. Although this species can be quite plentiful in certain areas, it never becomes a problem in the way the field vole sometimes does. Its main predators are the hen barrier and the short-eared owl.

The Orkney vole is confined to Orkney, where it occurs on all the larger islands except Hoy and Shapansay.

Bank Vole

Although it is less robustly built than the field vole—in profile more mouse-like—the bank vole reaches about the same size and weight: under five inches in length, with a tail of about two and a half inches. Tail length isn't much of a help if the vole is scurrying across your path; it is diagnostic if you have the animal in your hand. It is fully half the length of the body, whereas in the field vole it is well under half.

The upper fur of the bank vole is chestnut brown. The eyes are small and the ears partly hidden by the fur. The tail, which is often carried gay, is dark above and light below. Island races of this species are usually darker than the mainland ones, and are said to be better tempered.

Temper in small voles is a word to conjure with. It is usually said that, if you have a vole in hand, it's a bank vole if it bites and a field vole if it doesn't. This is, to a great extent, true. But field voles sometimes bite, maybe about one in ten of them, so you have to be careful that you don't pick up the tenth one.

Bank voles range high, and I've found them at over 1,800 feet in the Highlands. They have been recorded at 2,500 feet. They are most likely to be found in deciduous woodland, or in scrub where there is plenty of ground cover. Cover is what it likes best, so it will be found in hedgerows or banks where grass, brambles, dogrose and other trailing plants grow thickly. So long as there is this kind of cover, the bank vole will often be found far from woodland of any kind.

The nest of the bank vole is made of grass, moss and bark, all finely shredded, and may be built above or below ground. Nests above ground are built in thickets or mossed tree stumps. The breeding season begins in spring, and there may be four or five litters in the season. This high rate of increase is made possible by the fact that the vole can become pregnant while suckling, and despite the fact that few survive to breed for a second year.

The food of the bank vole is very much like that of the woodmouse, except that it takes fewer insects, or their larvae. They eat the leaves and stems of plants, besides a great variety of wild fruits, roots and fungi. Bank voles will sometimes eat field voles, a habit they share with the common shrew.

Although it is active at all hours, the bank vole has the eyes of the daylight forager, and it is by day that he is most likely to be out and about. He is preyed upon by birds like the kestrel, and by mammals like the weasel, stoat and fox. But bank vole doesn't figure as much in their prey list as one might think.

Compared with the field vole the bank vole is more active and agile, altogether more sprightly, a scurrier and a climber, a burrower, but not a user of grass creeps. He drives tunnels just under the surface, linked by runways in the grass, and deeper shafts, often under tree roots, to his sleeping quarters. The weasel is the only predator that can invade these burrows, but they are often used by other voles, mice and shrews.

The bank vole is found on the Highland mainland, but not on Orkney, Shetland, the Outer Hebrides and some of the Inner Hebrides. The bank voles of Mull and Raasay are considered as subspecies.

Opposite

The American Mink is now well established in many parts of the Highlands. It preys on small mammals, including rabbits, voles and mice. But it also takes birds, frogs newts and fish. The dark variety, known in the fur trade as standard, has the highest survival rate in the wild.

Left

Territorial Dog Weasel leaving his sleeping nest. Although able to kill prey much bigger than himself, the weasel usually stays well within his prey range, concentrating on small mammals. In young forests, the field vole is the main prey. A dog weasel requires about a vole a day; the bitch proportionally less.

Right

The Otter is now becoming a scarce animal over much of Britain but is still common in the West Highlands and Islands. It is still killed in many places because of its predation on fish. But, apart from fish ponds or fish farms, it is unlikely that otters make any impact at all on the fish population.

Below

Four-months old Badger cub calling to the sow badger at dusk. Cubs usually emerge from the set before the parents. At the age of four months, the cubs are self-supporting but there is still a strong family bond.

Pine Marten

Far left

Dog Stoat in February changing back to his brown summer coat. Highland stoats usually stay white longer than this. Some stoats make only a partial change to brown and white. Stoats that have changed completely to ermine pass through a brown and white stage when changing back to their summer coat. It is therefore difficult to tell, if one sees a brown and white stoat in February, whether it is an animal only partially changed or an ermine changing back.

Left

Adder casting its skin. Adders do this periodically, casting the entire skin, including the eye cover.

The Pine Marten, often called sweet marten, to distinguish it from the polecat (foul marten, foumart, or fitchet) is the graceful, agile weasel of the trees; the shy, elusive weasel of the ancient forest—where the giant capercaillie struts in the dawn: the hermit of wild gorges, where rowan and birch and aspen and drooping woodrush crowd down on rioting waters.

Once widely distributed in Britain, and a common species in the Highlands, the pine marten's range was severely restricted by the destruction of the forests, and it was thereafter cut to pieces by gin trapping. By the beginning of the twentieth century, it had been reduced to a remnant in north west Scotland. It is now recovering and colonising the new forests of sombre, regimented conifers. But it has no statutory protection. Its only sanctuaries are the Ben Eighe National Nature Reserve in Wester Ross and the goodwill of individual land-owners, including the Forestry Commission.

Behold the cat-size weasel with bushy tail and throat of creamy-white or orange. See him on a freezing winter morning, after a blizzard has blanketed the hills, with the white hares and the ptarmigan down, buzzards and kestrels on roadside posts, and grouse running like refugees across the ruffled snow by the roadside.

Small birds are fluttering and scolding in the roadside trees, in one of which is the marten—alert, with head up, tail down and rump arched. He leaps from branch to branch, climbs down backwards to the ground, and bounds across the snow to the next tree. His orange cravat is brilliant against the white. His bushy tail is twitching. He climbs the tree, curves round the trunk and leaps among the branches; then he descends at high speed and goes bounding uphill, slotting his footprints deep in the snow.

Or wait for one in midsummer, in the ancient pinewood, half an hour before sunset, overlooking a known trail. Nearby, is a pine marten den, recently used, but presently cold. A marten appears at darkening, proving for once that interception can be an art of the possible. He curves round the base of a tree, snake-like, white-fronted, bright-eyed, and stalks forward, padding low. He crosses our front and bounds uphill into the shadows of the forest.

The next night, he comes again, at the same time and by the same route. He twists round the tree, pads into the open, sniffs about, scratches himself, stalks towards us, turns uphill and disappears as before. On the third night, we move in closer to the trail, but the wind plays us false. At fifteen feet, the marten's nose takes in the man-smell, and he is away as though blasted from the spot—up through the forest, seeming to float over the shadowy ground before disappearing into a dark rock hole.

The pine marten is active at all hours, but probably hunts mostly from sunset till dawn. A Ross-shire farmer has seen them in his garden in daylight. My friend J. D. Lockie and I have over thirty such sight records: eleven by day, eight after dark, and twelve in the early morning and early evening. Some of the night sightings were in the headlamps of a car, when the animals were crossing the road into timber.

Like the stoat and weasel, the pine marten has good eyes and well-developed senses of smell and hearing; but his eyes are no more selective than theirs, and he can't recognise a motionless man merely by looking. He is a fast mover and a great jumper. A leap of six feet is commonplace, and a leap of more than twelve feet has been recorded down-hill. Like other weasels, he is usually a silent animal, but he can squeal when angry, and is said to purr and growl at mating time, like a badger.

In the Scottish Highlands, small rodents and birds make up the bulk of the pine marten's food during most of the year. The field vole is by far the commonest small mammal prey. Bird prey contains a high proportion of tits, wrens and treecreepers. Other items are hares and rabbits, beetles, caterpillars and cocoons, deer carrion, fish, wasp grubs and a variety of wild berries.

The marten is capable of killing poultry and game birds, and it was the game preserver who killed it almost out of existence. Man is still the main killer of martens in the Highlands, but much of the killing nowadays is accidental. The beast has no trap sense and is frequently caught in gins. Out of thirty-three martens examined after death, twenty-eight had been caught in gin traps. The gin became illegal on 1st April, 1973.

Like the stoat, the pine marten breeds only once a year, and there is delayed implantation. Mating time is July and August, but implantation does not take place until the middle of the following January. The young, averaging three in a litter, are born in March and April. They begin to appear at the den mouth at the age of eight weeks, and are probably independent at the end of five months. The young of both sexes reach adult weight in their first summer, but do not come into breeding condition until the following year.

Marten dens in the Highlands are most often in cairns, crags, or holes in trees; but breeding has been recorded in the old nest of a buzzard, in peat hags, and under the peat crown of a large boulder. But the beast has, on its range, a number of temporary dens or resting places that it uses from time to time, leaving signs of its presence. Nest boxes have been put out on Ben Eighe but, so far, no marten has bred in one. Other temporary resting places are the unused eyries of golden eagles and inside the moss cushions of old tree stumps.

The pine marten is big, ranging from two-and-a-half feet to two-feet-nine inches in overall length. The tail is ten or eleven inches. Males are bigger and heavier than females. Weight ranges from two to nearly three-and-a-half pounds, depending on sex, age and time of year.

The colour of the marten's coat varies from pale chestnut to near black, and there is a creamy-white or orange throat patch of variable size. The orange tint comes in with the winter coat in September and October. This coat is shed in the Spring and, by June, the animal has a thinner tail and the short, dark fur of summer.

In recent years, the pine marten has been increasing in the north west and spreading out from there into new forests, north and west of the Grampians where it is being regularly reported on new ground.

Stoat

Like the weasel, the stoat has suffered from its publicists. The general impression is of a blood-thirsty marauder, opening jugular veins, drinking spouting blood and terrorising the countryside round the clock. All sorts of pejorative names have been applied to it—cut-throat, gangster, wanton killer, murderer—none of which has any validity in zoology, or in a sane man's vocabulary.

The stoat as blood-sucker is a fantasy. No stoat sucks blood but all of them, like any other carnivore, will lick it. The idea that the animal is a combination of surgeon and vampire is nonsense. The stoat kills his prey by a bite on the neck, close to the ear, and this is where he licks. It is also nonsense to say that the stoat needs blood to slake his thirst. He drinks water.

The stoat is the only British member of the weasel family that turns white in winter. Then he becomes an ermine. The change to white is normal in the Highlands. Generally speaking, the further north one goes, the earlier the change to white and the longer the stoat stays white. The moult to white begins in November, and may be completed in seventy-two hours, or take several weeks. The change back to summer brown can be equally dramatic, or prolonged.

Stoats range high, and can be tracked at

3,000 feet even in mid-winter. The male stoat holds the territory and the female lives within it. But she does not appear to be so hemmed in as the bitch weasel. In fact, all observations indicate that dog and bitch stoat live amicably together at all times. One gets the impression that territorial frontiers become less well defined towards the high tops, where there might well be a sort of no man's land.

Stoats breed only once a year, the young being born in late April or early May. Mating takes place in July, but there is a long period of delayed implantation, lasting until March of the following year, when development of the young really begins. Young stoats, like young weasels, grow quickly, and are driven off the home territory when self-supporting. The female kits come into breeding condition in July, so are very possibly mated with the territorial male, who may or may not be their father.

The bitch stoat makes her nest in a rabbit burrow, a hole under the roots of a tree, or in a cavity in a bank or wall. If the site is dry, she lines it with only a little grass or other herbage; but some nests are heavily lined and the fur of prey is added as the days go by. Nearby, the stoat has her latrine. Long before their eyes open, the kits crawl from the nest and use this. As in the case of the weasel, heaps of droppings can betray the nest, and I found one on Mull in this way. When I looked into the burrow, the bitch stoat spat in my eye.

Although the stoat has the eyes of the night hunter, it is active, and sometimes hunts, by day. Daytime is often playtime, and there is no British animal, including the otter, more playful at any age. A family of stoats will play follow my leader in and out of an old wall, along the trunk of a windfall, or up and down a tree, chuckling and crooning as they chase each other.

A single stoat will play by himself, snake-walking and somersaulting, chasing his tail, leaping, rolling, bouncing and cart-wheeling. One winter day, I watched one on top of an old straining post, spinning as though on top of a gramophone turntable. Later, he hung from the top wire of the fence, upside-down and upside-up. He swung along it, hand over hand; he sloth-crawled underneath. He was a furred snake, spring-loaded.

The stoat is an excellent climber, a good swimmer and an expert tracker. When hunting above ground he frequently sits up tall to look and listen. When running a rabbit trail he keeps his nose to it with great single-mindedness, ignoring all other rabbits within sight or smell. His final approach to a stalked prey is often a bounding rush.

A rabbit being tracked by a stoat quickly crouches down and waits to be killed. If it is picked up before the stoat gets to it, it recovers from its traumatic shock.

Like the weasel, the stoat hunts underground, but he needs more elbow room. He will hunt a dry-stane dyke from end to end, following where any small rabbit can go. In the same way, he works field drains, rat holes and the burrows of water voles. The burrows of field voles are too small for him, but he can break into the grass creeps.

In young forest plantations, the main food is field voles; but the prey range is wide. When wild prey is abundant, the stoat can be choosey, leaving bits and pieces lying about. When he has to work harder for his living, he eats everything that is eatable. He will store surplus food and draw on his store if he has to; if he does not need it, he forgets about it.

When threatened, the stoat will face literally anything—dog, cat, fox and even man. That is, if he is cornered. If there is an escape route he will take it. He will bite, hiss and claw so long as there is a breath left in him. He will bite at the cat that has crippled him; put an inexperienced cat to flight; and snarl in the face of the dog that is breaking his back. Yet, in captivity, the stoat is much more withdrawn, and far less aggressive, than the smaller weasel.

When playing together, stoats croon and purr. An angry stoat will chatter. Young stoats have a bird-like call but, generally speaking, they are more silent than vocal.

Parties of stoats are sometimes reported in mid-winter, especially during prolonged snow. These are probably animals leaving an area where there has been a failure of food supply or some other upset. Many such parties must disperse without incident, but sometimes they get mixed up with man or dog and make the news. They may be families or aggregates of individuals. They are no particular menace to anybody.

Stoats are found over most of the Highland mainland, but are missing from the Outer Hebrides and most of the Inner Hebrides (Rum, Arran, Colonsay). They are found on the island of Mull, but not in Orkney. They have been introduced to the Shetland mainland. The stoats of Islay and Jura are smaller than the mainland ones, and are treated as a sub-species.

Stoat at Play
No Highland mammal, not even the otter, is more playful than the stoat. A bitch stoat playing with kits will croon to them, and the kits croon to her. This crooning may be more than a contact signal; it may express pleasure. Stoats playing in a farm outbuilding will climb over everything while roughing and tumbling with each other. Periodically, the family will become explosively active, and perform a sort of "wall of death" act round the building. Some stoats amuse themselves by leaping and bouncing against a tree trunk. A pair playing this game leap alternately, performing beautiful figures of eight as they leap against the tree and curve back to the ground. The stoat will climb the roughcast of a building to play in the guttering and the down-pipe. Others like running through drain pipes, playing "follow my leader".

Mink

The Mink that is rapidly colonising the Highlands is American, not European. Since 1929, it has been ranched on an increasing scale and today's feral population is made up of recent escapes, and the wild-bred descendants of earlier escapes. Official policy is still to exterminate all feral mink; but the American is trying hard to be Highland and is obviously succeeding.

On average, the mink is about the size of a polecat-ferret, varying in length from seventeen to twenty-six inches, including its tail of five to nine inches. Males are very much bigger than females. Depending on sex and age wild-bred mink weigh between one-and-a-quarter and two-and-a-quarter pounds. Ranch-bred animals can run to twice, and almost three times, these weights.

Most people see the wild mink dead in a trap. Being still rare, and shy as the otter, it isn't very often seen on foot; but its tracks can easily be found along the banks of rivers, streams, ponds and lakes, which are the main habitats of the wild population. It isn't so difficult to see a mink regularly over several days, once you know it is there.

Despite its reputation for savagery and shyness, the wild-caught mink quickly settles down in captivity, becoming friendly and feeding from the hand. I have three, wild-caught as adults, that are now perfectly tame.

There have been several accounts of confrontations between man and mink—some from the Highland mainland and one from the Outer Isles. Reports like this have suggested a fierce animal, prepared to attack man. Like any of the weasels, the mink will fight if it is left no way of escape, but the idea that it is a dangerous animal to meet is not borne out by the facts. Like all the weasels, it has an abundant curiosity and, if a man sits down to wait for a mink to reappear, it will almost certainly do so—asking questions rather than making threats. I have had a feral mink come up and take a bait from my fingers, leaving me with a completely digited hand.

Like the otter, the mink is a first-class swimmer, and it can catch frogs and small fishes. It climbs wire netting easily, and is agile when running along a windfall or up a sloping trunk. But it cannot compare with the stoat or the marten as a tree climber.

One usually hears about the mink after it has done some damage, such as raiding a hen-house. Then it makes the news. But most wild mink are never heard of because they are not getting into any obvious trouble.

The wild mink is rich, dark brown in colour, almost black, with a white spot on the chin and lower lip. Other colours turn up in wild populations, but whether they survive to breed is another matter. Most trapped animals are of the ancestral type.

Mink breed in the Spring, and there is delayed implantation. Litter size is usually five or six, but can be much higher. In Scotland, young feral mink have been recorded only in the Spring.

Not much is known about the prey of the mink in the Highlands. Fish and small mammals are taken, and it is likely that the prey list there must be as varied as in America. The mink may become a problem animal in the Highlands; or it may not. This remains to be seen. It may, on the other hand, fit nicely into the niche vacated by the polecat.

The distribution of the mink in the Highlands area is not known. One can say it is liable to be seen anywhere in the Highland mainland or in the Outer Isles.

There is no doubt at all that the mink is now thoroughly established in many parts of the Highlands, and breeding successfully. It is unlikely now that any campaign to exterminate it will succeed. There is now some interest, not to say concern, in its relationships with the otter. They share the same habitat some of the time.

Weasel

The weasel is a kind of Ishmael, with every man's hand against it, and there must be few keepers' vermin boards that don't have their array of corpses—shrivelled, eyeless, grimacing, moulting, peeling or flyblown—strung out like a game display in a poulterer's window.

Like so many mustelids this one has been badly served by its publicists. The tiniest of Highland carnivores, a mere fingerling, a pocket Nimrod, it is looked upon by many people as a kind of Bengal tiger, a furred Dracula ravening round the clock. I once cleared a pub of notable hard-men by depositing a 3-ounce tame bitch weasel on the bar.

Of course any weasel that is cornered, and left no escape route, will defend itself, using threat and explosive contumely. Even a tiny bitch can be a firecracker, swearing and sizzling almost eye to eye, and shrieking like a file on the teeth of a saw. If you insist on poking your nose into her breeding nest you're quite likely to pull back with her hanging on the end of it. So? After all the brooding dove will slap your face with a wing if you push it close enough to her nest.

There is still a widely held belief that there are two species of weasel in Britain: a big one and a small one. And there are big weasels and small ones. But only one species. The fact is that the big weasels are dogs, and the small ones bitches.

This disparity in size between the sexes is characteristic of many mammals, and is most striking in the weasel. Dogs are twice, and sometimes more than twice, the weight of bitches, and it is this dimorphism that has given rise to the belief in two species. In the canon of this faith the small females are called mouse weasels. The name has no validity. All weasels are mousers.

Highland dog weasels weigh over six, and up to seven, ounces, depending on the age of the animal and the time of year. Like everything else, weasels have their fat times and their lean times. Some dogs go well above seven ounces, and I weighed one that was over nine. Bitches range up to three ounces, or a little over. The biggest Highland dog weasels make the size of the average bitch stoat.

Sex for sex, of course, weasels are smaller than stoats, and the two need never be confused. Apart from a general similarity in shape and movements they are different in almost every other way. The weasel has a smooth-furred, two-inch tag of a tail, with no black tip. The fur is rich red-brown above and white below, with an oak-leaf line of demarcation. There are two brown spots under the chin, one on each side, that vary in size and site from weasel to weasel.

Like the stoat the weasel is territorial. The master dogs hold the ground, which they defend against other males of their own species. Stoats may be on the same ground, and in young coniferous plantations almost invariably are, but there is no territorial rivalry between the two species. The one doesn't exclude the other. Nor is there direct competition, although there may be argumentative confrontations. The weasel doesn't scare readily, and a big dog will easily drive a bitch stoat out of his way if he has any reason to.

Territory varies from 2 to 10 acres, and is marked by scent, urine and droppings. The bitch lives within the dog's territory, hunting her own range, but she has greater freedom of movement when she is nursing young. This is indicated by livetrapping marked animals, and the behaviour of captive weasels suggests that motherhood gives her a freedom that is lost once her family has been weaned.

The kit's eyes open at the age of 31 days, by which time they are chewing flesh. From the earliest days they crawl from the nest to defecate; later they come right outside to do so. A nest is therefore often betrayed by the heap of droppings on the doorstep. Fresh droppings are black and shiny, but soon dry off. They are the size and shape of gorse pods.

Breeding nests may be under tree roots or boulders, or in burrows or dry-stane dykes. The nest is made of hay, bitten off and carried home by the bitch. Later she may build the skins of voles and mice into it, so that it becomes a fur ball, sometimes as big as a wasp's bike.

Weasel kits grow rapidly, and with abundant food their growth rate is startling. By the age of 8 weeks the bitch kits are as big as their mother, and the dogs have far outgrown her. Then the dark brown fur of the kits is the only way of distinguishing them from their parents.

Once the kits are self supporting they are driven off the territory by the old dog, who may kill one or more of them if they take too long about it or get in his way too often. The bitch may then rear a second family, as many Lowland weasels do, but it isn't known if, or how often, this happens in the Highlands.

The breeding season is from March to September, first litters being born in April and May. In the best habitats, as in a young coniferous planting swarming with voles, double littering is more than likely. Litter size varies from 4 to 11, but 6 to 8 is more likely.

A nest of weasels, after the kit's eyes have opened, is like a hornet's nest—a dark snuggery of needle teeth instead of needles of flame and venom. The bitch will die in defence of her kits, facing man or dog if need be and, if they are handled, she will certainly move them, dragging them by the scruff of the neck at high speed, and not as gently as a cat carries kittens.

If you are on weasel territory, and sitting quietly, you may see the bitch on a foray with kits. If a kit is killed on a road the family will return and drag it into cover. This habit is exploited by keepers who will leave a dead weasel near a trap to attract others. This corpse summons seldom fails, and another weasel runs into one end of a tunnel but doesn't come out at the other.

Weasels don't range as high as stoats, and you won't find many pushing hard on 1,500 feet. They like lower ground with plenty of cover, and travel as much underground as aboveground, using mole runs and other hidden alleyways. From time to time they surface to sit tall, and look and listen.

They climb well and are good swimmers. Their prey range is narrow, and they like to keep well within it, but they can kill prey as big as, or bigger, than themselves when they have to. Bitch weasels can't handle big grandfather rats; dog weasels can. They take small birds and eggs, insects, small fishes and frogs. But their main prey is small mammals, and in young coniferous plantings these are almost entirely field voles. A pair of weasels, rearing two families of six kits, kill something like 2,000 voles in a year.

Weasels are found over most of the Highland mainland, in varying density, but are absent from all the islands except Skye.

Polecat

He came hurrying down through the barred and chequered moonlight of the wood—sinuous, darting, rippling; leaping from mossed stump to black-veined boulder; pattering along the brashed trunks of fresh windfalls—a giant weasel with yellow cheeks, pale lips and ears edged with white; masked and dark furred, with bottle brush tail. And, by all accounts, he had no right to be there.

But he *was* there. He was no wraith. However, being a polecat—called by some foul-marten, or foumart, or fitchet, to distinguish him from his relative, the pine-marten, or sweet-marten—there was doubt about his pedigree, because it is maintained that there are no true polecats left in Scotland. All Scottish polecats, it is argued, are mere feral polecat ferrets. This question of right polecats and wrong polecats, which seems to me to be what it amounts to, is worth looking at.

There may be a few polecats left in Caithness. If there are, they are likely to

Dog Polecat. Opposite
The status of the polecat in the Highlands is far from clear. Any polecat found there is usually written off as a polecat-ferret. But the polecats of the Island of Mull are as real as any polecats anywhere. This is a Mull polecat.

be the real thing. There may be some real ones surviving in the wilds of Sutherland. But there are definitely polecats on the Island of Mull about which there is no doubt whatever. They are the descendants of polecat ferrets introduced to the island a long time ago to kill rabbits. Their descendants are rejected by the systematists because they are descended from feral polecat ferrets. But it seems to me that a polecat that looks like a polecat, acts like a polecat, is accepted by other polecats as a polecat, and breeds polecats that look like polecats, is a polecat for all practical purposes. The argument becomes academic if the polecat happens to raid your henhouse. Real or *ersatz*, the result is the same.

In any part of the Highlands, one is liable to come across a polecat ferret running wild. This is especially true in areas where rabbits are numerous because it is in such places that the ferreter will be operating and, there, he is always liable to lose his worker.

In the polecat, the outer coat is long and dark, with a purplish sheen. The outer fur parts readily to reveal the buff under-fur. There is a mask of dark fur enclosing the eyes and extending over the muzzle. The mask is framed with creamy-white, often almost pure white, giving the impression of spectacles.

Dog polecats weigh up to $2\frac{3}{4}$ pounds and reach almost two feet in total length. About seven inches of this is tail. Bitch polecats are lighter and smaller.

The polecat occupies rocky hillsides, scrub areas and screes up to the 1,500 feet contour, and occurs most commonly in thick cover where it will den up in rabbit burrows, or under rocks or tree roots. Nobody knows whether the polecat breeds only once or twice in a year, and there is very little exact information from the Continent on this subject. It is thought that a female who loses her young will come on heat again and produce a second family. I know that Mull polecats can have a second litter in captivity. Four or five is the usual number. There is every reason to believe that the polecat and the ferret are alike in the matter of breeding. It would be surprising if they were not.

There is no doubt that, in the nineteenth century, the rabbit was a common prey of the polecat; but it killed then, as now, a great variety of prey, including game birds. It also takes eggs. It kills voles and mice and rats; fish, snakes and frogs.

Man was, and still is, the main predator on the polecat. The beast was originally killed for its skins, and it is difficult nowadays to realise how common the animal was up to the middle of the nineteenth century. For example, it is a matter of record that 3,000 polecats' skins were offered for sale at the Dumfries Fur Sale between 1830 and 1835. There was then a thriving trade in pelts. It was the gamekeeper and the game preserver who changed all this by classing the polecat as vermin and killing it hard by every means in their power. Being a great predator on the rabbit, the polecat was easily caught in the rabbit catchers' traps.

A feral, or wild-bred polecat ferret, may not be a real polecat in the eyes of the purist, but the real polecat is prepared to mate with such an animal, and the young are polecats for all practical purposes. The question of right and wrong polecats is one to which there is no easy answer.

Observe then a big wrong polecat living the free life and thriving on it. He is dark-furred, sheened like a sloe, with dark mask and spectacles of greyish-white. He comes from the glen, bounding from cover to cover, from hazel to birch, from birch to juniper—a gigantic weasel with hump up, moving fast.

On the edge of the forest, he sits tall, looking and listening, displaying chest and belly of black and bronze; then he drops to all fours and bounds purposefully into the timber.

By the rules, he is a wrong polecat, but he looks right. Although I don't know it at the time, he has come to stay a while. On that first day, a rabbit dies. I find it bitten on the neck and partly eaten. The wrong polecat returns to the prey and eats again, then enters the house and falls asleep. He has known people, so has to be a polecat ferret; but no wrong polecat ever looked as right as he did. He weighed 2½ pounds, not struggling.

Like all the weasels—except the otter, and maybe the badger—the polecat has had to wear many labels; blood-thirsty killer, terror of the woods, cut-throat, gangster, murderer, killing for the sake of killing. As usual, most of these accusations have very little relation to the truth. Like the others, the polecat had his niche. He killed to eat and took what was most readily available. During an abundance of prey species, such as rabbits or voles, he would take more than he could eat, eating only the choicest parts. This is characteristic of predators, including man, and has nothing to do with blood-lust—an emotive term invented by those who were hysterically opposed to anything that looked the wrong way at a game bird.

Unlike the pine marten, the polecat is a poor climber; but he can move fast on the ground and is a first-class swimmer. In this he comes closer to the mink than the marten. Like the other mustelines, he has a stink gland from which he sets scent to mark his territory, and it is the secretion from this gland that has given him his other names of foul marten and foumart.

It is recorded that young polecats (as I have observed with young weasels) learn the art of killing by trial and error. In the early days of their apprenticeship they often bite prey in the wrong place.

Polecats are night hunters but, like stoat, weasel and marten, may be seen hunting by day. If startled or frightened, they stink. They can cluck, chatter and yelp, but are usually silent. Wrong polecats are usually silent too, but have the same vocal range.

If the polecats of Mull are not classed as right ones, then there are no polecats in the Highlands and Islands. But if the wrong ones, so-called, are to be accepted, there are polecats all over the area.

Badger

Boar Badger. Opposite
Like the real tough guys among men, the badger does not go looking for fights. He is a stranger to bragadoccio. Although a confirmed non-starter of trouble, he is ever prepared for the brulzie if it comes his way, and is a terrible executioner at the finishing of it. Highland badgers are bigger than Lowland ones, which is true of carnivores in general. They increase in size northwards on their range. Although an adult boar badger is fit to cripple or kill any full-grown fox, it is a fact that badgers sometimes put up with foxes in their set. But, from what can be observed, the fox is confined to a part not in use by the badgers. Badger cubs and fox cubs can be seen outside the same set, but at different entrance holes. All the evidence suggests that there is no underground mixing, and one would imagine that if the vixen wandered through to an occupied part, she would be immediately thrown out if not physically assaulted.

The badger is bear-like, squat, big-clawed, hand-footed, hen-toed and out at elbow, with the bear's inswinging shambling gait. He is giant of frame, heavy of bone, broad of chest, sheathed in muscle, strong-jawed, powerful, retiring, tolerant, non-aggressive, and a terrible fighting machine, but he is the last only when self-effacement, turning the other cheek, and the peace conference have failed. Once he goes to war, he can break a terrier's jaw, cripple a collie, kill a fox, or take the hand off a man.

Bear-like though he may be in conformation, the brock is a weasel. The flick of the head may be bear; but once he starts moving he is all weasel, with the typical rippling hump-up run of the clan. In action he looks very like the wolverine.

Everywhere he goes, he is heralded by his white face. Watch him coming through the gloom of a birch-thicket in the early morning. Only the white of his face is visible. It weaves, and bobs, and flickers, disappears and reappears, following the ups and downs and windings of his homeward trail. It is a warning flash signalling forward what is following on. *Stand aside*, it flashes *Molest me, at your peril*! In moonlight or dark, or the half-light of the evening, his candy-striped face is as plain as the stripes on a skunk.

The brock is a beast of ancient lineage; his ancestors knew the cave bear. In modern times, he has had the experience of being evicted by mechanical earth-movers and buried by bulldozers. But he is a thrawn stay-at-home, reluctant to quit. His set is his castle, and his castle is his home. He puts the *Bydand* sign up, and defies everything and everybody.

The earth-works of an established badger-set are unmistakable: tons of earth fronting a varying number of deep burrows, so that the whole resembles a series of small quarries. Single holes and single mounds are just as easy to identify; size and depth give them away.

Badgers use bedding of hay and bracken. At an occupied set there will be, at different times, old bedding stuffed in a hole outside, bedding mixed with excavated earth, or trails showing where new bedding has been trundled in. When a badger is gathering bedding, it drags it backwards towards the set. But if it is gathering long hay, it might push it ahead rather than drag it.

Though badgers will range far in a night, they will be home in the morning before daylight, if not in the main set, then in a spare one nearby, unless they have moved in to spend the day with neighbours further out.

The brock is a clean housekeeper, and you will never find a set cluttered up with rubbish as you will a fox den. If it is cluttered up outside, you can be sure there is a fox lodging inside. Vixens sometimes whelp in a badger set, keeping to a single-end out of the badger's way. The brocks will usually put up with a fox lodger, but sometimes they take a turavee and throw the lodger out. A badger can, and at times does, kill fox cubs. But such killings must be rare. I once knew a boar badger that killed a vixen in a gin trap outside his den; then he was caught in a second trap himself.

Badgers don't carry food home or to the den-mouth. It follows from this that if lambs or poultry are found at a badger den, the culprit is a fox lodger. I have used the word "never" here—always a dangerous word to use where animals are concerned—but, in this context, it is justified. I know of no-one who has ever seen a badger carry food into a set.

Small prey rather than large is the badger's way—rats, mice, voles, rabbits, slugs, earthworms, caterpillars. He takes a great variety of wild fruits, including dog-hips, also a variety of roots, bulbs and tubers. He will eat grass. Gamekeepers often accuse the badger of being a considerable predator on game birds and their eggs, but there is no evidence of this beyond their say-so. Evidence from dung and stomach analysis is mostly negative. Then there are wasps. Badgers delight in tearing out wasps' nests and devouring the grubs. They will wreck any nest they

come across. They also like the grubs and honey of the bumble-bee.

When rabbits were plentiful, badgers dug out a lot of nests and one could find evidence of this all over a wood. As a general rule, the badger digs down on to the nest from above—not in from the entrance. In this he differs from the fox, who digs in from the entrance.

In the badger, there is delayed implantation. The main mating period is spring. Implantation takes place in December to January, after which development is complete in about two months. Cubs are born in February and March, and began to appear at the den-mouth when they are about two months old, although they are active below ground before then.

At three months of age, they will still accept one's foot as a normal part of the scenery, and run over it again and again during play. They are extremely playful at this age and can make as much noise as a litter of piglets in the bracken and undergrowth. As June wears on, they become more concerned about foraging and usually leave the set soon after emerging. They are often out in good light and on their way before the adults appear.

Badgers aren't often seen about by day but the spring mating period is a likely time. This is when I have seen badgers on foot in the Highlands, among rocks and bracken. Cubs are more prone to come out exploring in daylight. Some Highland badgers have their young above ground among rocks or bracken and such cubs are readily put on foot by day.

A badger's nose is good, very good, and can't be fooled. His hearing is acute but his eyesight isn't as good as yours or mine in the dark. A sniff of you, however slight, or a peep out of you, will send him rumbling down into his den; but he will look you in the face without flinching so long as you are not actually recognisable for what you are. You will see a badger when he won't see you.

Like the other weasels and the fox, the badger has musk glands with which he sets scent. You can watch them doing this. It is certainly an advertisement, and probably marks territory. It enables badgers to find their way and each other. Sows set scent for their cubs, as you can see by the way the cubs follow the mother's line some time after she has left on her rounds.

Badgers don't hibernate although they are less active in winter. Snow on the ground doesn't keep them in, not even in the Highlands. December is a quiet month. January is often the opposite. A February freeze-up doesn't keep them in much, presumably because they have cubs, or are about to have them. I have tracked badgers in February snow right over the 1,500 feet contour in different parts of the Highlands.

Badgers that die below ground are buried there by the others, walled off in a blind-end burrow. This is borne out by the fact that one finds skulls and other bones in new diggings from time to time.

Badger runways are clearly defined when in nightly use and can easily be mistaken for footpaths until they pass under some obstacle where a human being couldn't go unless he was twelve inches tall. Then the mistake becomes apparent. The brocks use these routes to and from the den as people use pavements.

Apart from the expected snorts, snuffles, grunts, squeaks and yelps one would expect from such an animal, the badger is normally very silent. The boar purrs to the sow when he is calling her out. Then there is the badger's scream. Nobody knows why a badger should scream as it sometimes does.

In this country, the badger has no enemies except man and his dogs. Terriers put to ground after a fox sometimes come up against a badger and may be severely mauled. The man who deliberately sends a terrier to ground to face a badger should resign from the human race. Such dogs are often chopped, chewed up and literally painted with their own blood.

Otter

An Otter whistles at dusk, then falls silent, and there is only the whisper of the riverside trees and the faint whooping of peewits tossing about the sky over the hill. The trees' reflections, ragged of edge, lie darkly on the pool with a lane of sky between, and there is a greyness of foam where a log is snagged among roots below the opposite bank. The roots reach down to the water like arthritic fingers and probe under the bank where their tips are buried: the part of the bank that covered them has long since been eroded away.

A ripple spreads out from the darkness under the far bank and presently a shape swims noiselessly into the sky lane—a small shape trailing a V-wake. The shape turns in the sky lane, ringing the water, and follows it across the pool. A rat? A water-vole? No: it is an otter—swimming with only her flat head showing. Now the paw-prints on the silt, and the glazed slipway into the pool are explained.

The otter up-ends and dives, then breaks surface buoyantly like a seal—without betraying splash, but in a vortex of ripples. She swims across the lane and back, turning on her side, half rolling, curling like a caterpillar, sinuous and silent. She trails her wake into the near shadows, and is invisible. But there is splash and disturbance there, so perhaps she has cleared the water to snatch at trailing catkins. When she reappears in the lane, she follows it upstream across the pool, into the dark shallows, and is lost. And from there comes the clear whistle of an otter, answered from downstream.

It is April, the time of leafing, and the smalltalk of swallows is in the air. Woodpigeons are clattering into the oaks on the slope and croodling in the pines across the river. The daylight is slow and grey, with the tree-tops stirring. This morning, the river song is clink and pebble chatter, and tits are belling in the alders and the bank.

The flat-topped boulder in mid-stream is this morning a plinth for a sculpture—the ebony silhouette of an otter, round-rumped, with head high and rudder hanging slack. For long seconds, she holds this pose; then she pads round in a complete circle and sits tall like the giant weasel she is. She settles again, and trails her dark, sleek length across the boulder, with her face close to the water on one side, and her rudder clasping down on the other. She holds this pose for a few moments before curving, snake-like, from the boulder into the water—disappearing as though she were gliding into a burrow.

Although highly specialised for his life in the water, the otter has not become an over-specialised hunter. He can become a land-weasel when he has to, or wants to. But it is as a fisherman that we see him in his proper role, whether he is killing eels, coarse fish, frogs or salmon. He takes what is to hand, and eels make up a large part of his diet. He has, therefore, his basic needs, and a river that can't support him in the way of fish will seldom hold him for long, no matter how many rabbits or other prey there may be around about.

The otter is known as a great wanderer, and the reason for these wanderings is not yet fully understood. The impression one gets on the rocky coasts of Wester Ross and the Western Isles is that otters are evenly distributed at intervals of several miles along the best river habitats. If, in fact, the otter holds a territory, like the stoat, the weasel and the pine marten, wandering otters might well be dispersing juveniles, or nomadic animals, in search of a range of their own.

Whatever the explanation there is the fact that otters are great wanderers, travelling down to the sea, from sea to head-waters, and from one water-shed to another. They can be tracked in snow through the contours, from glen to glen, up to 1,200 feet above sea level. They will cross wide moorlands and other open country by lochans, ponds and streams. They may desert a stretch of river for a while, and fail to settle in a habitat that appears to have everything they require. There is no doubt that the strongholds of the otter today are in the West Highlands and Islands where many of them haunt the shore-line. The animals will hunt the kingly salmon—and some of the finest salmon rivers in the world are in the West—but it is unlikely that otters make any more impression on the numbers of salmon than I do.

The otter has a gland under his tail from which he sets scent on logs, branches, boulders, stones and tufts of vegetation along the river-bank. This, combined with his habit of sprainting (defecating) in prominent places, suggests territorial behaviour and a method of contact between individuals. The wandering otter, indeed, follows the trails of long dead generations of his kind, using the same land routes and water routes and crossing places, and for the same reasons.

In winter, the otter is betrayed by his tracks more obviously than in summer. These, and the furrow made by his body, can be followed in the snow. He plays in snow as joyfully as he plays in the water, burrowing in it, rolling in it, and sliding down the river-bank. Such snow-slides are easily spotted. Otter tracks on ice often lead to where the beast has rolled, or skidded, or both, apparently for the joy of it.

The most settled time for otters is when they are breeding, for the dog will remain with the bitch at least until after her cubs are weaned, and the bitch won't move on until they are swimming strongly and catching their own food, unless danger forces her to carry them or lead them elsewhere. She may stay with them right into the winter. The dog otter usually lies downstream from the holt and will fight with any other dog otter who tries to move up past him. It is often taken for granted that the dog otter develops a wanderlust after the cubs are weaned, but this is by no means certain.

The otter breeds in any month of the year, and there is practically no evidence indicating a peak period. The usual litter is two or three cubs, but up to five have been found. The breeding holt is usually away from the main stream and, as often as not, has an entrance under water. Others are heavily screened by washed-out tree roots or other vegetation. Some breeding nests are in woodland or thickets. Sea-caves and rock holes are used by marine otters, not to be confused with the Sea Otter which is a separate species. Otter nests are usually made of reeds, moss and grass. Just as often, the cubs are found lying on the bare ground, like fox cubs.

Otter cubs do not swim readily, but they don't have to be taught to swim. They don't swim until they have grown their waterproof coat at the age of three months. Then they have to be enticed or forced into the water by the bitch, who tends them there. They are reluctant swimmers, but the ability to swim is inborn. By the bitch's example, and with her help, they proceed to develop their full powers and learn the act of hunting fish.

Otters are found from the headwaters of rivers down to the estuaries and the sea; in lowland lochs and in mountain tarns. They live by rocky shores and in sea-caves. Clean water, salt or fresh, is the otter's hunting ground and its holt is never very far away from it.

In many ways, the otter resembles a seal. It has the seal's streamlining; the powerful nose and ear muscles that snap shut before a dive; and the heavy bush of sensitive whiskers set in a swollen upper lip. It is, in fact, a web-footed weasel, with a thick, tapering tail, supremely adapted for its dual role of land and water hunter.

Dog otters weigh between twenty and twenty-five pounds, and beasts heavier than that are common enough. Bitch otters weigh from fifteen to eighteen pounds. Dogs measure about four feet two inches, including tail; bitches about three and a half feet. In both sexes, the tail is more than half the length of the body.

Wildcat

See him—the Scottish tiger—padding across a wild scree in the slanting morning light—big-fisted, wide-eared, long of limb and tusk, with ringed club-tail—and you have seen one of the lords of life: a cateran of fire and brimstone, implacably savage, reputedly untameable. . . .

Cry Vermin! if you like. Throw up the gun or wave out the dogs. Or set the steel-toothed trap on the fallen birch that spans the burn, so that he'll be taken by a foot when he walks across. Then you have a dead cat, with the fire extinct, and the weapons of war perished. But see him at bay on a ledge of some rugged crag, after a run before terriers—flat-eared and bristling, with yellow, moon-eyes slitted, teeth bared to the gums, back arched and tail bottle-brushed—and you have seen explosive cattiness in all its demonic fury.

Hark to the wild pibroch of him—the eldritch laughter, hiss and crackle, scream and sob—as he stalks through the gloom of the corrie when the moon is riding high! The wildcat's skelloch is of the lonely places, of the high forests and mountain fastnesses—haunt of the mountain fox and eagle, the red deer and the ptarmigan—although in recent years he has been skirmishing along the frontier of civilisation, with the bright lights in his eyes.

You will hear the treason whispered that there's no such animal; that there isn't one of the entire clan that hasn't a drop of domestic blood somewhere in its pedigree. The drop may be there—which is by no means certain—but there's no doubt at all that the wildcat of today is all the cat he ever was, as strong and robust as his prehistoric ancestors.

There should never be any real difficulty in telling the wildcat from any gone-wild domestic type. Apart from size and weaponry, the wildcat's club tail, clearly ringed in black, is diagnostic. All domestic cats have tapering tails. Any big cat seen on the hill after dark, or caught in car headlights on a Highland road, is more likely to be a true wildcat than a domestic one.

Kitten of the Scottish Wildcat
The kitten of the Scottish Wildcat looks very much like the striped kitten of a domestic cat, and is just as playful until approached by a human being, when it displays explosively, baring its teeth and striking with its claws. The wildcat is said to be untameable but if kittens are taken before their eyes have opened or just as they are opening, they can be tamed in the usual way. Even older kittens can be tamed if one spends enough time in the effort.

It is a great pity that most people's impression of the Highland wildcat is of a bristling, snarling fury, with teeth bared and claws unsheathed. The impression derives mostly from photographs showing a cat at bay; and it is false. The wildcat is no more a habitual snarler or spitter than the domestic cat, and will always avoid man when it can.

It has to be emphasized that a gone-wild domestic cat is not a wildcat, and never can become one, no matter how long it lives the free life. The true wildcat is a species—not a way of life.

In general, the wildcat is bigger and heavier than the domestic cat. A mature male will measure over three feet from nose to tail tip, the body being fully two feet long. Weight ranges from seven pounds to fourteen pounds or more, males being heavier than females.

In the Highlands, female wildcats come into breeding condition in the first half of March. The gestation period is sixty-three days. The den is usually in a cairn, or under the roots of a tree but, occasionally, the cat will use the old eyrie of a golden eagle. The usual litter is two or three. The kittens are reared by the female and hidden away from the tom who might kill them. At the age of four or five weeks, they begin to play about the den mouth and, soon afterwards, they follow their mother on hunting expeditions. Like the mountain fox, the wildcat will readily move her kits to a new den.

It isn't clear whether the wildcat breeds more than once in a year. Kittens have been found from mid-April to the end of August, and some are born as late as December. So it may well be that the wildcat, like the otter, will breed in almost any month of the year.

The prey of the Highland wildcat ranges from grasshoppers to roe deer fawns. Between these extremes of its prey range, it takes voles, small birds, fish, red grouse, black grouse, hares, rabbits and even capercaillies. A charge of lamb killing is occasionally made against it, which may or may not be true. Examination of wildcat droppings in Wester Ross showed that the main food was vole and rabbit.

The wildcat hunts at dusk and dawn, and during the night. But it will also hunt by day, especially when the nights are short or during a lean spell in mid-winter. I have watched one crossing a ridge at over 1,000 feet in deep snow at mid-day when the sun was shining. A female with well grown kits is a likely one to see by day in a quiet place.

I had my longest view of a free-moving adult wildcat from a ledge beside a golden eagle's eyrie where I was perched in a camouflaged hiding place. It was early morning, with low mists, and the sun coming up in crimson and orange flames. The wildcat came padding down through the rocks into the corrie, with myself at house-height above her and the eagle on the nest paying attention. The cat passed within twenty feet of the base of the eagle's rock, stopped, looked back, scratched herself, then padded on: bush-tailed, moon-faced, with out-pointing ears.

The wildcat's eyes and ears are good, and it uses both when stalking prey. Its nose is not so good. Like the domestic cat, it will wait on for the reappearance of a quarry that has escaped into a burrow.

Unlike the domestic cat, it does not cover its droppings.

The wild tom holds and defends a territory and marks it with his droppings and urine. He has places where he sharpens his claws—trees or scrub or rocks, as the case may be, and he may set scent at the same time from the glands in his feet.

Since the middle of the 19th Century, Britain has had no wildcats outside the Scottish Highlands. At the beginning of the 20th Century, the species was at a dangerously low level and threatened with extinction. Then it was found in only the most remote cairns of the north west. Now, it is found in all the Highland counties and is still spreading.

Opposite

The Wildcat is now more numerous than it was fifty years ago, and is steadily colonising new ground.

Left

Short-tailed or Field Vole
This is the rodent found on thick grasslands and in young forest plantations. In such places, its numbers fluctuate on a four year cycle.

Right

The Brown Rat is a ubiquitous species found from sea level to mountain tops and on small islands. It is preyed upon by anything able to catch and kill it.

Below

The Wood Mouse, also known as the long-tailed field mouse, is a common animal in woodland and on grassland. It is strictly nocturnal and has the dark eyes of the night hunter. It is even inhibited by strong moonlight.

Red Squirrel

The Highlands are still the stronghold of the Red Squirrel, which has been disappearing from many other parts of Britain. It is a species of coniferous forests, where it lives largely on the seeds of cones. Some red squirrels also take the eggs and young of small birds. Squirrels can very readily be brought to a bait of hazel nuts, raisins or chocolate and soon become very tame where they are not molested.

The Red Squirrel sits on a high pine branch, close to the trunk, twirling a pine-cone with his fore-paws while he strips it. The scales and a few golden seeds fall on the snow-drift below, freckling it with red. The last of the snow lies banked like a parapet along the length of the ride, under the trees, and the white has a rash of cone-scales dropped by other squirrels. It is a day of ice and icicles, blue sky and frost, but there is no wind to stir the dark pine-tops.

Presently, the squirrel drops the core of his cone, which is golden with a club-end and twin green needles attached to the club. It falls down the snow parapet and comes to rest on the slope. The squirrel, hand-footed, now runs along the branch, which bends under his weight as he nears the tip. There he selects another cone which he deftly snips off with his teeth and carries to his perch near the trunk. He sits up, facing the ride and, holding the cone in his fore-paws, begins to flake it with his chisel teeth.

In a tree further down the ride, two other squirrels are perched, flaking cones and showering the snow-bank with scales. Scales and cores lie close-ringed round a stump at the end of the snow-bank where another squirrel, breaking the rules, has been feeding on the ground.

The first squirrel now comes head-first down the trunk. After pausing about six feet from the ground to look and listen, alert and bright-eyed, he leaps down, skips airily over the snow parapet, crosses the narrow ride, and climbs into a pine tree on the other side, rattling and rustling among the twig litter and dead sticks as he ascends. He swings in the crown, shaking the needled clubs, then leaps into the next tree, and the next, until all sight and movement of him is lost. He is off to his bed.

On the far side of the wood, there is another ride, a wide one with massed pines on one side and a ridge of hardwoods on the other. Here squirrels also feed, sometimes in the pines and sometimes on the ground underneath the hardwoods, where the leaf-litter is deep. This morning a breath of frost lies on it, and the iron is in the ground, but a squirrel is working there.

He is working the slope, keeping close to the base of a massive oak. He is hopping about, or scurrying in short spurts, with his brush-tail following, not arched. Every now and again he stops to scrape in the leaf-litter. Each time he finds something he sits up, with tail arched, and eats it from his fore-paws, rejecting the bits he doesn't want.

It is impossible to guess whether the squirrel is seeking seeds he has hidden there and remembered about, or finding seeds he has hidden and forgotten about, or whether he is merely seeking on chance. But he is certainly finding something, for he sits up at short intervals with forepaws held to his face and teeth gnawing. Some of his finds he carries to an exposed root of the oak tree on which he sits, clear of the ground litter to eat.

When his hunger is satisfied, he crosses to the pine-wood, but not by way of the ground. He climbs into the oak, stopping every few feet to look down over his shoulder, with tail hanging and forepaws gripping wide; then he runs along a high branch and leaps out and downwards into the top of the nearest pine. It is a sort of parachuting act, with legs widespread and tail bushed out behind. There is a shaking in the pine-top and he has disappeared.

At the same time of the year, many miles further north, where the pine forests flank the territory of the eagle and the wildcat, the red deer and the mountain fox, the squirrels are out, feeding high in tall larches. The drifts are six feet deep below. The sky is freezing blue and the winter sun hardly warms, but each morning and late afternoon the squirrels can be seen high in the larches. Out on the hill the stoats and the mountain hares are white. The ptarmigan are still on the white tops, and the deer are down in the glens. Yet the legend of squirrel hibernation persists.

Squirrels lay on fat in the autumn and draw on their store during the winter; but there is no hibernation. Cold means nothing to them. Sleep means no more to a squirrel than it does to a badger or a man. It is a resting period between spells of activity. The idea of the torpid squirrel is wrong.

During the very worst of the weather—snow, high winds, blizzards, driving sleet or rain—he will lie up in his nest for a day or two, but he will go out as soon as he can because he has to eat. Where the squirrel is lying up like this, he probably dozes most of the time, like a dog kept indoors when there is nothing else for it to do. You won't catch a squirrel in a deep sleep, as you will a dormouse or a hedgehog. He may sleep snugly through the worst storms but he is only sleeping.

Red squirrels hide surplus food. Many other mammals, and some birds, do the same, and remember where they have hidden it. The squirrel hides surplus food, mainly by burying it, but sometimes by hiding it in holes in trees. Many more, perhaps most, bury their finds in caches all over the place. One thing is certain; squirrels don't find everything they hide.

Squirrels in coniferous woodland or forest spend very little time on the ground, which is not surprising in a true tree squirrel. The red squirrel is born in a tree, lives in trees, lives off trees, sleeps and nests in trees, and is seldom seen far away from trees. When surprised on the ground, his escape route is up into a tree. Squirrels have even climbed telegraph poles in a panic. There is a story recorded by Barret Hamilton of a squirrel that climbed a man. Confronted by a Highlander and his dog on a treeless moor the squirrel climbed the man to escape from the dog.

In the trees, squirrels are safe from practically everything except a man with a gun. On the ground they are more vulnerable. I have known a stoat kill a squirrel.

The drey of the red squirrel is built usually more than twenty feet from the ground, in a fork against the main stem of the tree, and is bigger than a football. It is made of twigs, and lined with moss, leaves, grass and bits of bark. There is no obvious entrance or exit, so the squirrel has to force an entry at each visit. Once she has her young in the nest, the female will not allow the male into it, so he sleeps in another one near at hand.

When the female comes to have her second litter, she builds a new nest near the first one, so several dreys can be found in a very small area.

Young red squirrels are suckled for about seven weeks, by which time they are running about outside the nest and nibbling at buds, green shoots and bark. They become more and more venturesome with each day that passes and soon make their first journeys to the ground, scampering into the tree again at the first hint of danger.

Besides being ornamental, the squirrel's tail is a muff that he wraps round himself when asleep. It is a rudder and balancing pole when he is swimming, leaping or climbing, and acts as a kind of parachute when he makes a long jump from one tree to another or from a branch to the ground. The hairs of the tail are under control so he can make them lie flat or brush them out. The movement of the hair betrays the squirrel's reactions, as a roe-deer's white rump hairs betray his.

The red squirrel was an extremely common animal all over the Highlands before the eighteenth century but the destruction of the forests in the eighteenth and nineteenth centuries reduced its numbers drastically. It may even have become extinct. Red Squirrels from England were introduced into Scotland, and quickly built up their numbers because their introduction coincided with the maturing of new forests. It is very possible therefore, that all our Highland squirrels are Sassenachs. The ancestral home of the red squirrel is the dark coniferous forests, so it is still most numerous where the old forests remain or where new ones have been planted as, for example, on Speyside.

Grey Squirrel

This is a North American species, first introduced to England in 1876. Thereafter there were other introductions, and it was deliberately spread in Britain by man. It was still being planted around as late as 1929, and 10 years later had reached Loch Long in the west and Fife in the East. It is now at the Highland frontier, and may be across it.

The grey is a big, strong squirrel (11½ inches plus tail against the red squirrel's 8½ inches plus tail) sometimes called the tree rat. It is a hustler and a coloniser, ecologically fitted for a hardwood forest niche. It found in England hardwood niches of the kind that suited it, and in such places it ousted the native red, for which such habitats are secondary. In direct confrontation with the red (not the main factor in its success) its strength carried it through. Sometimes it killed red squirrels—mainly young, but adults too. It remains to be seen if it can colonise the great coniferous forests of the Highlands, which are red squirrel strongholds.

Acorns and beech mast are main food items, but the grey's taste is catholic, and it eats all hardwood seeds and shoots; roots, bark, bulbs and catkins; fungus, grain, eggs and flesh. It buries surplus food. It strips the bark off hardwood trees, frequently killing them, and is now a major pest against which all-out war is being waged. It is increasing in numbers, and still spreading. There is some evidence of natural population declines, and recoveries, related to changing food supply; there is also a little of emigration. It is illegal to keep a grey squirrel for any purpose without a special permit from the Ministry of Agriculture.

The grey squirrel breeds twice a year, building its drey up to forty feet from the ground. It is most numerous in broad-leaved or mixed woodland, but will nest in conifers where there are enough seed-bearing hardwoods close by. It is not found in big coniferous forests yet.

Wood Mouse

Woodmouse
The Woodmouse used to be more commonly called the fieldmouse. It is a nocturnal animal, with dark eyes with the eye-shine of the night hunter. It is put off even by bright moonlight. It does not hibernate, but stores food for the winter, sometimes in underground burrows, sometimes in old hedgerow nests. It is a spritely animal, able to jump like a flea.

The Wood Mouse is very like the House Mouse but it has longer legs, longer ears, much bigger eyes, and a more massive snout. It is a brown mouse, more reddish on the rump, and shading to buff on the nape and head. The underside is white or grey-white and there is a flush of yellow where the two meet. On the throat there is often a patch of orange which sometimes becomes a collar or extends as a chest stripe. The tail is extremely long and will be over three and a half inches in a mouse measuring seven and a half inches overall. Fully grown adults weigh from half an ounce to an ounce but none ever reaches the weight of the St. Kilda race which can reach almost two ounces.

As its name suggests, the wood mouse is most commonly found in woodland and scrub areas. It is also common in the hedgerows of cultivated fields. It is a hardy animal and will be found in small numbers in open hill country wherever there is suitable food and some kind of cover. It ranges widely over some of the treeless Hebridean Islands.

The breeding season is from April to October and in some years it goes on longer, even through the winter. The mouse builds her nest of chopped and shredded grasses either below ground or under a heap of branches. Underground nests are at the end of a short burrow. The mouse's permanent nest may be as deep as three feet.

Wood mice eat a wide range of foods, including insects and their larvae, snails, spiders, seeds, nuts, fruits, buds, green plants and fungi. They are occasional predators on the eggs and young of small birds. The mouse, in turn, is preyed upon by owls and weasels. Foxes kill them. Country cats also take their share.

The most outstanding feature of the wood mouse, apart from its long tail and boot-button eyes, is its spriteliness. It can climb up, down and around, move with bewildering speed, and jump like a flea. Despite this agility, the vertical flea-jump does not save the wood mouse from the fox, which can snap it out of the air as easily as a terrier can snatch a fly.

In his underground citadel the wood mouse is safe from all mammal hunters except an invading weasel. Being mainly a night prowler, he is not troubled by hawks or kestrels but the moment he appears from his burrow he is liable to be attacked by cat, fox, weasel or owl.

Wood mice store food, usually underground but sometimes in an old bird's nest in a hedge. Some wood mice will roof over such an old nest and use it as a resting place as well as a food store. Very often an old bird's nest is nothing more than a seat for the mouse when it is eating, and the litter from such meals can be found scattered on the ground below. Underground nests often have food in them; seeds, oats and haws.

Wood mice climb into trees and hedges to collect hips, haws and rowans. They reject the pulp and extract the seed by nibbling a hole in the same way as they nibble a hole in hazel-nuts to extract the kernel. They tackle the hard shell at the most convenient place. They do the same with the shells of snails.

Wood mice squeak but are not notably noisy; nor are they aggressively territorial. Individuals and family groups seem to live amicably together.

The wood mouse is found all over the Highland mainland but is rare on high mountains and heather moorland. It is found on many of the islands, even those without trees, but is absent from North Rona and Lunga of the Treshnish Isles. There are three Island races, all with sub-specific status—the Hebridean, the Shetland and the one on the main Island of the St. Kilda group.

House Mouse

The House Mouse has been here since before recorded history, so it is a truly Gaelic species despite its Far Eastern origin. House mice living in the Far East exist out of doors and feed on the seeds of grass, but as soon as the animal associates with man its feeding habits change and it develops new types with longer tails and darker bodies. Such are the mice we have with us. Fraser Darling has told how, when he was living on Lunga of the Treshnish Isles, the house mice moved into his camp, although they had been living wild for eighty years.

House mice living in corn ricks spill out into the surrounding countryside, so the corn ricks become a sort of recruiting depot for those living in the field. The mice living indoors or in corn ricks probably breed throughout the year. Those living out of doors probably stop breeding during the winter. A female will produce from five to ten litters per year depending on the quality of the habitat. Usual litter size is five or six young. House mice build their nests of any suitable material—paper, wool, straw, hay, cotton, nylon or anything that is warm or can be shredded. In corn ricks breeding nests are often communal.

Little comes amiss to the house mouse in the way of food. He eats seeds, grain and all kinds of foodstuffs. Mice living out of doors probably eat the same diet as wood mice.

Cats and Barn Owls are the main predators of the house mouse. Weasels and stoats can work great execution in a corn rick. Elsewhere the mice are taken by whatever predator can catch them.

The house mouse is found over the Highland mainland and on most of the Islands. The St. Kilda mouse was larger and darker than the mainland type but is now extinct.

Brown Rat

The Brown Rat is tough, resourceful, cunning and aggressive; bold and shy, wary and reckless; wasteful and provident, clean and dirty; highly adaptable yet hide-bound; quick to learn and slow to forget. As a result, rats have built up a vast army of occupation in this country, including the Highlands, and become the most costly and serious mammal pest in our history.

A big rat will face man, dog, cat or weasel when cornered. A big rat, especially a female defending young, will sometimes face man, dog, cat or weasel when not cornered. She will put a kitten to flight, maul a kit weasel, make a hole in a dog's face, or bite a man. A big buck rat, disabled or cornered, will do the same, and may do so at times when not disabled or cornered. He will show fight if he even thinks his escape route is cut off.

Although wasteful and destructive, the rat is also a food hoarder. I've watched them in a stable carrying away a pound of oats in the early part of one night. I once put down eggs, weighing $2\frac{1}{4}$ ounces each, and the rats rolled them away, using nose and chin, and heaving them over obstacles with their forepaws. One rat, reaching down from a hole six inches above the floor, contrived to lift an egg and drag it out of sight. I have seen a rat killing fortnight-old pullets by tearing them on the throat and neck, and have watched one manipulating eggs from under a broody hen without disturbing the bird.

The brown rat is found from sea level to the tops of mountains, on the seashore, and in the heart of the hill country, in houses and ditches, and along river-banks. It is a supremely adaptable animal, and may make seasonal changes of habitat—out of buildings to the rivers and hill in the summer, and back indoors in winter.

Under the most favourable conditions, brown rats will breed right through the year, but they will breed for only part of the year when living a harder life afield. Even when there is breeding throughout

The Woodmouse is nocturnal and will shun even bright moonlight so it isn't often seen alive and free. It will come into houses but is not the mouse found in corn ricks.

the year, not all the females in a population are breeding at the same time, or all the time.

The female rat makes her nest of whatever warm material is handy—rags, paper, straw, hay, binder twine. If unduly disturbed, she will carry off very small young and lead bigger ones away. I have seen two females in the same nest—a nest as big as a hat-box—containing fifteen young. Rats in outbuildings can do execution among nesting swallows. They will try hard to reach eggs or nestlings, but a strange object, placed on a rafter, will often keep them away for days at a time.

Rats, it is well-known, react strongly to the presence of new objects on their familiar territory—hence their wariness about poison baits, cage traps and spring traps. Having their regular runways, and being familiar with every yard of them, they will pull up short of any strange object placed in their path. A rat that used the same rafter nightly was never seen to do so again after a ·22 bullet had been fired at it.

Not all rats live in close association with man. Some live with him all the time, some of them some of the time, and some of them none of the time. Indoor rats regularly move out to rabbit-warrens and water-vole burrows in summer. But, at all times, there are outliers, preying on whatever they can catch, as well as eating vegetable matter and carrion.

Rats swim, climb and go anywhere feet can go. They will raid woodpigeon nests in oaks and thorns, and small birds' nests on the ground. On the Island of Eigg, they raid the shearwater's burrows. On Rum, they appear under any deer carcass left ungathered for a few days. They are opportunist and resolute. They will find what is to be found, and go where they want to go.

On uninhabited islands, where there are no human beings to plunder, the rats can exist indefinitely on crabs, fish, shell-fish and dead birds. Where the island has a colony of nesting sea-birds, the rats will eat the eggs and chicks in the breeding season. I saw an eyrie on the Island of Lewis in which rats were a regular prey. They were being caught by the eagle on an offshore island where the rats were living the life of beachcombers and thriving on it.

The brown rat has many enemies—man, dogs, cats, stoats, weasels, foxes, otters, badgers, eagles, buzzards, herons and owls. Many dogs, but especially terriers, are natural rat hunters, and will kill large numbers at one time for sport, whereas a cat will be content with killing one which it may, or may not, eat. Wild predators that kill rats eat them.

Traditionally, the farm cat is the supreme predator on rats, but cats cannot control or exterminate a rat population. The cat's nature is to take what it wants; not to go to war. Charles Elton has said that cats can keep a place clear of rats once it has been cleared, but this is not always so. I know of one place—indeed I have lived in one—where there was no history of rats, but where rats moved in despite the presence of cats and a terrier. The new factor was the arrival of a variety of livestock and the consequent large stores of foodstuffs.

Outlying rats, living in river-banks or in hedge-banks, are often seen on foot by day, but in farms and buildings they are active mostly after nightfall and before daylight. The signs of them are plain—trackways, tracks, droppings, holes, tunnels and places soiled by the constant rubbing of their bodies.

Right

Male and female Frogs in amplexus. There is dimorphism in frogs—the males being smaller than the females. This is clearly indicated in the photograph. The male frog holds his position by means of the suction pads on his wrists and does not release the female until she has spawned.

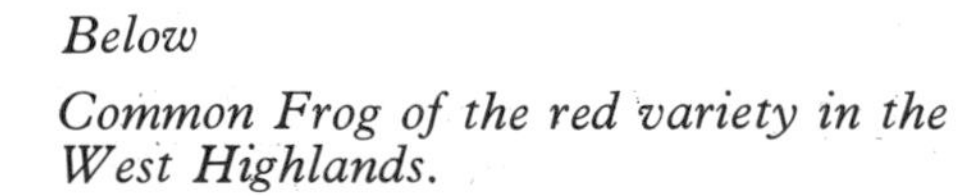

Below

Common Frog of the red variety in the West Highlands.

Left

Female Adders are, generally, less contrastingly coloured than males. They are also bigger. Adders are not aggressive snakes but they are highly excitable. The proper way to handle an adder, male or female, is by the tip of the tail. The snake should be held at arm's length, when it can do no harm.

Right

Slow Worms, also known as Blind Worms, are completely harmless. They feed mainly on small slugs, tame readily, and make attractive pets.

Far right

The Common Seal, unlike the Atlantic or grey seal, does not gather in large numbers in the breeding season. Common seal pups are born all round the coast and are able to swim from birth.

Grey Seal

The Grey Seal is far and away the biggest wild animal to be seen in the Highlands—or in Britain—the bulls being more than three times the weight of a good Highland stag. It is also one of the most predictable because its breeding stations are known and it is land-based until the breeding season.

In early Autumn, the seals begin to move towards their breeding areas, in twos and threes and small groups. They vanish from the coastal waters where they have been seen on and off all summer. They come in from the open sea: great, sleek, mature streamlined bulls, pregnant cows and seals of lesser degree—driven by a common, irresistible impulse and navigating with inborn infallibility towards their known landfall.

From the grey North Sea or the incredible greens and peacock blues of Hebridean waters they steer their way. They drive for the treacherous tideways of the Orkneys, to the remote Shetlands, and to storm-lashed North Rona—Island of the Seals and the biggest breeding station in the world—where rats long ago exterminated the people by consuming their food, and the last woman was found dead on the rocks with her child at her breast.

Perhaps each drives towards its own—the place where it was born. Thus the cow seal will return to her birthplace to give birth to her pup. This is possible for it has been noted that there are slight differences in seal colour, in behaviour, and breeding times between one colony and another. They are like Clans with their separate tartans. Whether or not each colony is a kind of closed society, the movement of the seals in space and time is the same.

But the incoming seals do not arrive like an invading army, or massed starling flocks flighting to a winter roost; they come in groups and waves over a period, following advanced parties of mature bulls. Before moving into the breeding grounds, the seals assemble on the perimeter, or on skerries close at hand. They haul out and lie there, the sexes usually separate—the bulls mostly inactive, the cows noisy and inclined to be quarrelsome. Yearlings haul out on other rocks. When this resting period is over, the seals invade the breeding ground and the season really begins.

On North Rona, many bulls move inland, some to the 300 feet contour, heaving themselves uphill on their powerful fore-flippers and with much straining of muscles like a man doing press-up exercises on the floor.

Once he has established his territory—on land, in a pool or offshore—the bull defends it actively, mainly by threat, but sometimes by actual fighting. Serious clashes are rare, but they do occur. The giant, ponderous bulls, displaying remarkable agility, bite and slash at each other like fighting pigs, ripping each other's hides and sometimes leaving gaping wounds. The cows move quietly ashore and gather round the spaced-out bulls.

But there is no strict harem system. The bull defends his territory while the cow holds an area round her pup. She has considerable freedom of movement and may mate with any bull she chooses once she is ready.

At the beginning of the season, the bull may have as many as fifteen cows on his territory. Later, he may have less than half that number. It is unlikely that the same bull can last out the two months of the season, any more than a red stag can last out the rut, as fresh animals are always challenging him and cows keep coming into season. The cows mate after weaning their pups and return at once to the sea where they begin to feed after their three weeks' fast. The bulls may fast for the whole two months of the breeding season.

As the season advances, the ground begins to smell of death and decay—rotting seaweed, urine, afterbirths and dead pups. Losses among pups can be

This pup is a moulter in its sea-going coat.

high on crowded ground, and as many as one out of five may die from starvation or injury as a result of squabbles among adults. Others develop septic sores that clear up if the victims survive to reach the healing sea. Yet others born on exposed ledges or platforms are swept away by storm waves and pounded to death among the rocks. A very young pup falling into the sea is in real difficulty if the cow is not at hand to rescue it.

At birth, the seal pup smells faintly of the sea. It weighs from 20 to 42 pounds, but grows rapidly because of the richness of seal milk and puts on weight at the rate of 4 pounds a day. Thus it reaches 70 to 100 pounds at the end of a fortnight. The first teeth are shed before birth; the adult teeth appear a day after. The pups are weaned at the age of 3 weeks.

Deserted by the cows at the age of three weeks, the pups begin to move about the colony, losing a pound of weight each day until they take to the water and learn to fend for themselves. A pup that goes to the sea without a good layer of blubber is at a great disadvantage.

During the breeding season, as indeed at any time, the grey seal is vocal, and it isn't difficult to understand the origin of the Gaelic legends of the seal woman, or singing seals, or other forms of anthropomorphism. The full range of voice can be heard on the breeding grounds—the snarls and bawls of the pups, the hooting and singing of the cows, and the hissing and grunting of the bulls.

Seal cows have no fear of a man in a boat. They will elbow their way down a beach and take to the water when they see a boat approaching, then bob up around it shortly afterwards, their round heads breaking the surface like so many divers' helmets. They will tread water, swoosh around, dive and reappear quite fearlessly, staring curiously with their gentle, water-washed eyes.

When hunting, the grey seal usually eats small prey under water. Large fish are brought to the surface, beheaded and skinned with strokes of the fore-flippers. The seal then lies back in the water and swallows the prey without effort.

Common Seal

Unlike Grey Seals, Common Seals don't come ashore to give birth or to mate; so there is no great seasonal shoreward movement or land-based assembly, and no territorial behaviour by the bulls.

Unlike young grey seals, which are helpless at birth, young common seals are active and able to swim. They are born either in the water or on a rock from which they float away on the first tide. The pup is tended constantly by its mother who rolls and plays with it in the water, dives with it and flippers it along in rough seas or through heavy swells.

This is the seal of coastal waters. In Summer, when the pups are being born, there is some concentration in certain favoured areas as, for example, round the Shetlands and the Hebrides. At other times, the common seal is found right round the Scottish coasts.

Characteristic of this seal is its water play in the month or so before the pups are born. The seals roll about, and porpoise, and slap the water—a display that is a delight to watch, whether in some busy sea lane or in blue Hebridean waters.

Common seal pups usually shed their white coats before birth but, now and again, one is born with it, and sheds it a day or two later. The mother suckles her pup under water for part of the suckling period, perhaps for all of it.

Bull and cow common seals can be distinguished from adult grey seals at all times by their snub noses and slanting nostrils. This seal is usually dark grey above and light grey on the underside, with darker spotting all over. Pups resemble the adults except for the few that are born white. Other names for the Common Seal are Harbour Seal and Brown Seal.

Like the grey seal, the common seal is a fish-eater. It has no Close Season. Fishermen kill common seals because they consider them serious competitors for flat fish. But there has never been a public outcry against this as there has been over the grey seal cull, perhaps because common seals are killed individually over a wide area throughout the year, whereas the grey seal pups are all killed within a very short period when they are massed helplessly ashore.

Pup of the Grey Seal
The dark mark on the shoulder is dye, daubed there for identification purposes during a field study. Adult seals are tagged, and the tag remains with them for many years. Seal pups, which shed their coats before going to sea, are dyed. The dye is lost once the pup moults. The dying enables the researcher to identify individual seal pups on a beach or an island.

Wild Goat

Goats have a tendency to run wild and have been doing so from the time they were first domesticated. The wild goats of the Highlands are the descendants of domestic animals that became feral a very long time ago.

The Persian Wild Goat is the ancestor of the domestic breeds, and all feral animals revert, in a few generations, to this ancestral type. Feral goats in the Highlands may be black-and-white, brown-and-white, black, brown or white, or even cream. Kids are often beautifully coloured. Adult billies grow horns like the true wild goat.

Where they are not being disturbed or shot at, wild goats can become very tolerant of human beings—down to a range of about a hundred yards. But on ground where they are really wild, they are as unapproachable and shy as the red deer, and extremely difficult to stalk.

When alarmed, goats give an explosive hiss that can be heard a long way off. When red deer hear this hiss, they know what it is all about and react accordingly. Thus, the wild goat can act as the ears and eyes of red deer before the red deer has time to use its own.

Goats move about in small herds and travel up and down through the contours. They can use the highest ground and survive in the toughest conditions, and will cling to the wild screes and the high tops when under pressure. Each herd is led by a master billy with a great sweep of horns and a good beard, and a smell that comes strongly on the wind. You can pack home part of a red deer without smelling of deer; but you won't pack home part of a wild billy without smelling of goat for long enough afterwards.

Feral goats eat grass and the coarsest vegetation on the hill. They strip the bark from branches and bushes, and from the trunks of hardwood trees. They can become a serious problem to the forester. In many places, they have been killed out. In most, they are tolerated; in some they are looked upon as an asset to the region.

The mating season is in October and November, when there is some fighting among the billies. They rear, come down, lower their heads and charge. Head to head collisions are common but appear to do little damage. A nanny goat with a kid will beat off the billy, engaging him horn to horn although hers are mere spikes compared with his.

Kids are born in the period February/March, when the weather can be most severe. Early birthdays, combined with a shriving environment, add up to rigorous selection, and mortality among kids is high. In addition to this kind of mortality, there is also some predation by eagles, mountain foxes and wildcats. The goat kid is well within the prey range of the fox and the eagle for about the same time as the black-faced lamb. But nanny goats can be bold in defence of their young and some are prepared to assault fox or dog on sight.

During the first week or two of its life, the kid stays close to its mother. Then it begins to play with other kids. In almost every way, the behaviour of goat kids is like that of lambs of the same age.

Wild goats are still widely distributed in the Highlands although they have disappeared from some places in recent years. In certain areas, they have been exterminated by the Forestry Commission. The map shows their present distribution in the Highlands and Islands.

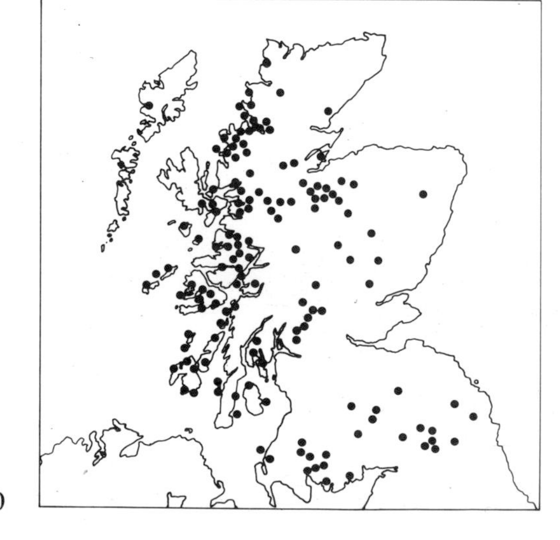

Soay Sheep

The Soay Sheep closely resembles the wild moufflon, but is smaller. It comes in two colour phases—the common dark brown type, which is almost black in autumn but lighter in spring, and the less common light brown type. Lambs vary in colour from black through dark brown to ginger. The adult fleece is a fine soft wool mixed with hair. Adults have the throat-hair and the mane of the wild moufflon. Soay lambs have the wild moufflon's horns, with a good rise and a full curve.

The tupping season is October, and the lambs are born in March and April. Many ewes become pregnant in their first winter, and many ram lambs of the year will breed in their first October.

The food of the Soay is grass and maritime vegetation.

Soay sheep cannot be herded by orthodox dogging. They scatter and run for their refuges on the cliffs. Dr. Morton Boyd tells how a good mainland sheepdog, working for a scientific party on Hirta, completely failed to herd them. The St. Kildans trained their own dogs to run individual sheep to a dead-end and hold them there. Most of the present-day soays run at the sight of a man, but the animals grazing about the old village have become familiar with the new residents and tolerate them to within ten yards.

Since 1955, the Hirta flocks have been counted each year. Dr. Boyd's census shows population fluctuation on a definite cycle. Each year, about one in ten of the population has been caught, measured and tagged.

The Soay sheep has inhabited St. Kilda with man for a thousand years and more, and may even have been there before the Viking conquest. They remain pure despite the fact that the St. Kildans introduced a number of breeds and crosses over the centuries. Segregated on Soay they survived as a type and no sheep quite like them exists today. In 1932, it was introduced to Hirta.

Although specimens of Soay sheep can be seen as park animals, and small units have now been established on other islands, the home of this ancient breed is Soay, Dun and Hirta of St. Kilda.

Wild Goats are widely distributed in the Highlands, and are found even in the Outer Hebrides. They are all the descendants of feral goats, but many herds are of great antiquity.

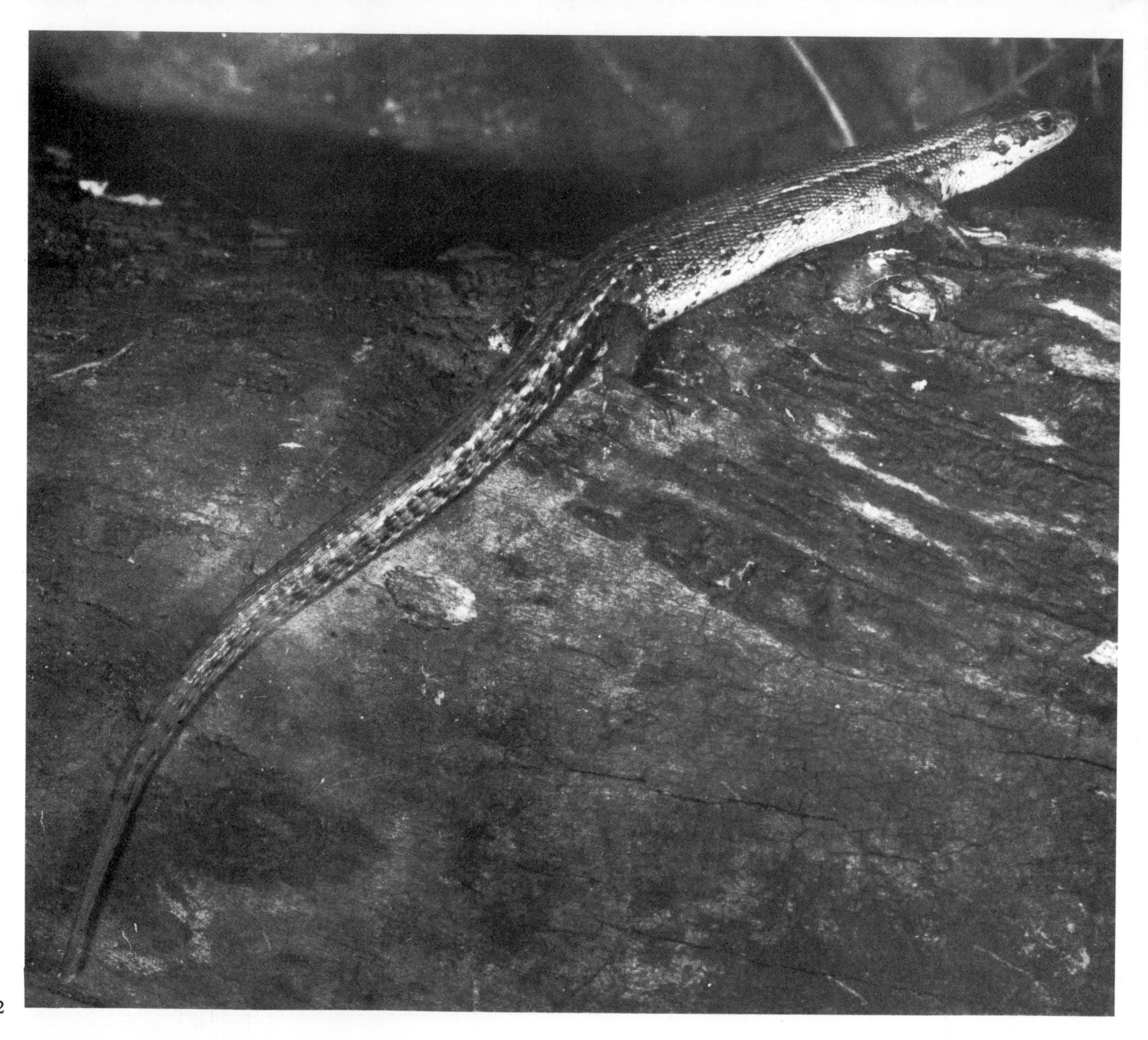

Adder

The Viviparous Lizard will be found among rocks and heather, especially in damp situations. It likes the sun, and is active in sunny weather. It knows every square foot of its small territory but, when frightened off it, it panics and is easily caught. It hibernates from October. If roughly handled this lizard will part with its tail. It may do so even when severely frightened.

There is only one snake in the Highlands—the adder; and it is poisonous. Therefore all Highland snakes are poisonous.

In many parts of the Highlands you'll come across a notice reading: Beware Adders. There may or may not be adders in the neighbourhood. Such notices are sometimes put there to keep away tourists. But whether there are adders on the ground or not there isn't much need to beware. The snake prefers to escape, undisturbed. Adders, as a rule, are in more danger from people than people are from adders. It would make more sense to put up a notice reading: Adders Beware of People. Except that adders can't read.

Most people are afraid of adders. Some run away from them; others react by staying to beat the snake to death with stick or stone. Neither action makes sense. All the adder wants is elbow room to get out of your way.

There is very little danger from adder bite in this country. Most people travel well shod, so most bites are likely to be on hand or finger. This is nothing to become alarmed about. I have been bitten three times in my lifetime. The first time I died was in 1929; the second time in 1960; and the last in 1970. I felt worse many a time when I was a callow Hogmanayer.

The important thing not to do when you are bitten by an adder is to run about in panic. You aren't likely to die unless you run yourself into the ground and die of exhaustion. You may be sick or headachy but you are most unlikely to die.

The adder has a small head, a narrow neck, and a thick muscular body, with the V-sign to the fore and a dark zig-zag line along its spine. These can be faint in dark females, but they can still be seen; in the more contrasty males they are obvious. Males are always more contrastingly patterned than females, and by this you shall know them.

Female adders reach two feet in length, or a bit over; males are almost invariably under two feet. Any adder will try to avoid you at first meeting, but if it coils up facing you the best thing to do is coil down beside it and speak to it. Being deaf, it won't hear you; but your immobility will pacify it.

Being a highly nervous snake, but not aggressive, the adder will likely be explosive on first acquaintance, but it is readily put at ease if you have the confidence to put it at ease. Many of them, in fact, soon reach the point where they make no attempt to strike. The safe way to handle an adder is by the tail, holding the snake at arm's length.

Adders hibernate from October until the spring, in clumps of brushwood, or under tree roots, or in holes in the ground, and they have a tendency to overwinter in groups. If you happen to be on the spot at waking time you may find up to a dozen of them together, coiled in the same brushwood or whatever. Picking one out of such a fankle, a man is most likely to be bitten. I was first bitten doing just that.

Soon after they emerge the adders pair, and the breeding month is May. If you see two adders at this time, reared and facing each other, you will be looking at a territorial battle between males and not a courtship dance as was once thought.

The young adders are born in early autumn, usually in August. They are pencil length, and not much thicker than corn stalks. They are boldly patterned, and obviously adders. They are venomous from the beginning but feed on small prey, usually insects and the like. Although independent from the beginning there is some association between them and the female, and they can be found in the same small area for many days. On hot days they are likely to be found under stones and boulders. There is no truth in the belief that the female will swallow her brood at the threat of danger.

The adder gives birth to fully developed young, the egg sacs rupturing at the moment of birth.

Like other snakes and lizards the adder has a forked tongue. This is a sense organ. The adder gathers scent impressions with it, which are transmitted to the brain. When it strikes a prey, say a mouse, the mouse is not killed at once. As often as not it escapes into cover and dies there. The adder tracks it to the spot with its tongue.

Adders prey on lizards, mice and voles. In turn they are preyed upon by some birds of prey. The hedgehog is a sometimes predator. But man is the main enemy of the adder in the Highlands, and most people kill it on sight.

The adder is found throughout the Highland mainland and on some of the islands. It is not found on Orkney, Shetland or the Outer Hebrides.

Common or Viviparous Lizard

There are two species of lizard in the Highlands, as easily told apart as a poacher and a salmon. The one has four legs: the other has none. The legless one is the slow-worm. The legged one is the common, or viviparous, lizard.

The common lizard is small—fully grown Highland adults measuring up to 140 millimetres in length, including tail. The colour of the upper parts varies a great deal—greyish, yellowish, gold, brown, olive—and is no guide to sex: the colour of the belly is. A lizard with a brilliant orange or vermillion belly, heavily spotted with black, will certainly be a male. One with belly of pale orange, yellow, grey or blue, with few or no black spots, will almost certainly be a female.

But to see this you have to handle the lizard, which has to be done with care, unless you want to be left with a tail while the tail-less front end scurries away. If you take hold of a lizard by the tail it will almost certainly leave you with it. It may even do so if you give it a big enough fright.

It isn't the handling that breaks the tail. Self-fracture, known as autotony, is under the control of the lizard. The bones of the tail are specially designed for easy fracture. As a protective mechanism it often works; the lizard escapes while a predator is left with the tail. The lizard goes on to grow a new tail which is never as good as the original and is made of gristle instead of bone.

This lizard, as the name viviparous indicates, gives birth to fully developed young, the egg sacs rupturing at the moment of birth or very soon afterwards, although in some cases there may be a lapse of a day or two. But the lizards inside these sacs are perfectly developed replicas of the adults.

The breeding season is in April and May, and the young are born in July, August and September. The female lizard prepares a cavity for her young, and prefers moist situations to dry ones. The average litter is five to eight. The young, at birth, measure up to 47 millimetres in overall length, the tail being less than half of this. They are dark—bronze or black—so can be easily distinguished from adults. They grow quickly during the summer and almost double their size before going into hibernation. Hibernation is from October until the Spring.

The common lizard feeds on spiders, insects, larvae, flies, grasshoppers and other small invertebrates. Some lizards eat earthworms; others do not. Small prey is swallowed whole but bigger prey is shaken in the way a terrier shakes a rat. The shaking stuns the prey, which is then chewed. The inside is swallowed and the skin left. In its turn the lizard is preyed upon by adders, crows, foxes, buzzards and others, but it is so active and agile that predators probably lose more than they catch.

This is an active lizard, able to climb walls and the smoothest of rocks with the greatest of ease. It moves in darts and scurries. It swims well and readily. It has its favourite basking places where you will see it by day, but it can't stand great heat. When it is basking on a stone or a boulder or a bare patch on a peat-bank, it can be watched without any difficulty.

You will find the common lizard from the lowest ground up to the 3,000 feet contour. In any situation it likes plenty of cover: among rocks, in peat-banks, under tree roots, on heather knolls.

The common lizard is found all over the Highland mainland and some of the Inner Hebrides, including Skye. It is absent from Orkney, Shetland, the Outer Isles, Mull and Tiree.

Slow-worm

The slow-worm is a legless lizard with a snake-like body, and is often erroneously called a grass-snake. It is totally unlike the adder, being smooth-scaled with a highly sheened skin that looks as though it had been varnished. Colour varies greatly—bronze, brass, chestnut, brown, grey, near-gold and even copper.

Like other lizards, and unlike the adder, the slow-worm can blink. It has no obvious neck. Like the adder, it casts its skin in one piece. It moves like the adder. Like the adder it picks up scent particles with its tongue. Unlike the adder it drinks with its tongue instead of thrusting its head into the water and gulping. It rarely bites and, when it does, the bite is completely harmless.

When first handled, the slow-worm will often struggle violently but it soon quietens down and becomes quite tame. It isn't as prone as the common lizard to part with its tail, even when excited. Nevertheless a great many slow-worms can be found with replacement tails, so they must lose them in some way. Males fight a great deal and many tails must be lost as a result.

The mating season is April to June; the young are born in August and September. Like the common lizard the slow-worm is viviparous, most of the egg sacs rupturing immediately. Newly born slow-worms are silver, gold or yellow above, with black flanks and belly. They become more uniformly coloured as they grow. At birth they are three to four inches long. They double this size by the end of the year, and grow half as long again after another year. They grow on for several years, so there is great variation in size among slow-worms in the same area. Fully grown Highland adults are around fifteen inches, more or less, but one finds bigger ones from time to time.

The slow-worm feeds on a variety of slugs, spiders, earthworms, insects and their larvae. It is very fond of the small white slug, *Agrolemix agrestis*, which it chews from end to end. Tame slow-worms will feed almost entirely on this slug, and can be kept in good health with very little other food.

Many things prey on the slow-worm—adders, buzzards, eagles, foxes, badgers, hedgehogs, crows, ravens. Even a greater variety of predators attack the young. Anything used to taking earthworms is liable to take a young slow-worm because it is about the right size and moves in much the same way.

The slow-worm is fond of dry ground and is not found in marshes; nor does it range very high. It swims well but quickly drowns if it can't find an easy way out of the water. It is a poor climber. When handled, it sometimes defecates.

Although it can be seen often enough during the day, the slow-worm isn't often found basking in the sun. Females are especially fond of lying in a warm place in spring and late summer. Most slow-worms like to be under cover. They are great burrowers, and are fond of lying under stones. In fact, the man who goes along turning rocks and stones is more likely to find slow-worms than the man who doesn't. Some slow-worms like to lie in loose earth, with only their heads showing. This happens regularly in newly worked-over soil in a garden. Other favourite places are sand-heaps.

The slow-worm is found throughout the Highland mainland except the North Eastern tip. It is found in the Outer and Inner Hebrides but not in Orkney or Shetland.

Common Frog

Opposite

The Common Toad is land-based outside the spawning season. Although people often confuse toads with frogs, this should never happen. The toad has a warty skin and crawls instead of leaping. A toad from very dirty or stagnant water may appear black. When clean, he is dressed in leopard-skin. Toads often come into outbuildings, potting-sheds and such places, and are easily tamed. They can be seen in the same place day after day.

In the Highlands, the common frog ranges up to 3,000 feet, and spawns up to 1,500 feet—in pools, ditches and puddles. I have found tadpoles at 1,500 feet in a temporary puddle on a deer path. The puddle dried out twice to thin mud, but the tadpoles survived. At 1,500 feet spawning is late because of the cold, but frogs in hill country prefer to go up some distance rather than stay in the glen bottom. Frogs several hundred feet up generally spawn earlier than those in the glen.

The frogs make their way to the spawning places in spring. It is not known how they find their way there, but it is increasingly thought that they do so by scent, recognising the smell of their own pond by the essential oils produced by the plants in it. The actual time of spawning is determined by the state of the algal growth. The frogs may mark time for a spell, waiting for the algae to come on, so that the tadpoles' food will be available when they hatch. This marking time period is known as the pre-spawning period.

At spawning time the male frog clasps the female round the neck, held in position by the suction pads on his hands. This position is known as amplexus. When she produces her spawn he fertilises it in the water. The spawn absorbs the water and becomes the jelly-like mass so familiar to every schoolboy.

After spawning the frogs disperse into the countryside, and this is when they may be found on the high tops among the deer and the ptarmigan. The females leave the pond before the males, so there are times when all the frogs in the pool may be males.

The Highlands are noted for their red frogs and I have certainly seen more red frogs in the West than anywhere else. But there is great variation in colour anywhere, and frogs in the Highlands come in all shades of yellow and sulphur to green and near-black.

Almost anything that eats anything will eat frog. I have found frog in the eyrie of the golden eagle and in the nest of a buzzard. I have seen ravens and crows bring in frog. Frogs are killed by stoats, mink, otters, polecats, wildcats, foxes, badgers, pine martens, herons, ducks, hedgehogs and rats.

The common frog is found throughout the Highland mainland and on all the islands of the West except the Outer Hebrides. It is absent from Orkney and Shetland. It was once introduced into Orkney but is presumably extinct there. More recently it has been introduced to Shetland.

Toad

Compared with the nervous, high-jumping, long-jumping frog, the toad is a phlegmatic crawler. It is squat, warty and wide-mouthed, with topaz eyes. Most toads are some shade of brown with grey or olive overtones. The handsomest look as though they were clad in leopard skin.

Frogs and toads are regularly found spawning in the same ponds or pools. Where this happens the two usually stay apart, the toads spawning in one part and the frogs in another. The female toad moves as she spawns, trailing her long strands behind her, and winding them round water plants. The eggs are dotted along the strands of jelly like dots on dominoes. After spawning the toads leave the water. They are much more land animals than frogs, much more given to settling in one place for long periods. They are also very long livers.

The secretions from the toads' warts are some form of defence against its enemies. These secretions gave rise to the notion

Crested Newt. Right
This species lays its eggs one at a time, and rolls them up in the leaves of water plants.

Smooth Newt. Far right
The Smooth Newt also lays its eggs on water plants but is not so careful about rolling them up. The eggs may even be laid in small clusters wherever there is a fork in the stem.

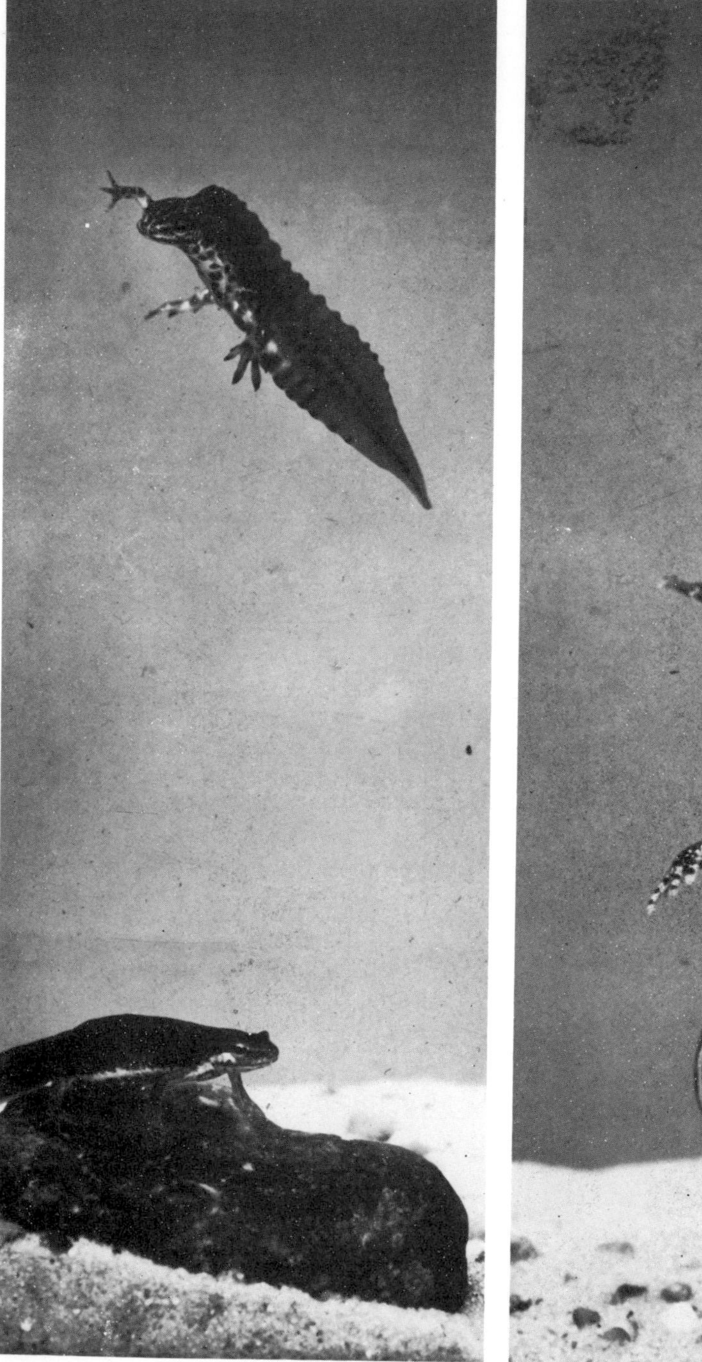

that the toad was poisonous. The secretion is, in fact, toxic. Despite this protection, many predators attack the toad, and some of them eat it. I have offered dead toad to hedgehogs, stoats and weasels but none of them would eat it. Rats, however, do kill them, and eat them after skinning them. Crows and ravens will kill them, usually opening them along the belly and taking only the inside.

So far as its own food is concerned, the toad will literally eat anything it can swallow. It eats all kinds of insects and their larvae, earthworms and wood-lice, slugs and snails. It will swallow newly-born slow-worms or even adders. It will sit by a wasp-nest and gather in wasps with its tongue. It does the same at a bee-hive. It eats a great many ants and, when in the water, will take young newts, small frogs and even small toads. The hunting toad isn't interested in anything that doesn't move. The prey has to move before it is recognised.

Like frogs, toads return to the same pond year after year for spawning. They hibernate on land—from October or November until the very early Spring. Favourite hibernation sites are dry bankings, or under rocks, or under tree roots. In more civilised surroundings they will use flower-pots in a greenhouse.

The toad is widely distributed over the Highland mainland. It is found in the Inner Hebrides, with the exception of Tiree, but is absent from the Outer Isles. It is found in Orkney, but not Shetland.

Newts

All three British newts are found in the Highland area.

The biggest, the Warty-newt, is recognised in spring because of its high crest, which goes from its eyes along the back to near the tail. The tail is also crested. After the breeding season, the crest disappears and becomes a very shallow fin.

In the breeding season frost and ice do not affect the newts. Egg-laying begins in April and continues into July, each female laying between 200 and 300 eggs. As soon as the tadpoles have become newts they leave the water and do not return there until they themselves are ready to breed. Because of the secretions of its skin, the warty-newt is not much bothered by predators.

The warty-newt is found on the Highland mainland, with the exception of the Caithness tip, and most of Argyll except the Mull of Kintyre. It is not found in the Outer Isles, the Inner Hebrides, Orkney or Shetland.

Smooth-Newt

The Smooth-newt is smaller than the warty. The male, in the breeding season, has a high crest from head to tail. The female has a low ridge of skin down her back. After the breeding season, the male's crest is much reduced and becomes a fin. The breeding male is olive green or brownish on the upper parts with a speckling of dark green or black. The female is brownish, olive or yellow, speckled with dark brown.

The Palmate-newt is the smallest of the three. It is very like the smooth-newt. The most obvious distinguishing feature is the dorsal crest which is shallower than that of the smooth-newt's and has a straight, not a wavy, edge.

The Palmate-newt is found over the whole of the Highland mainland, on Skye, and in parts of Argyll. It is not found in the Outer Hebrides, the Orkneys or the Shetlands.

Smooth-newt—see map.

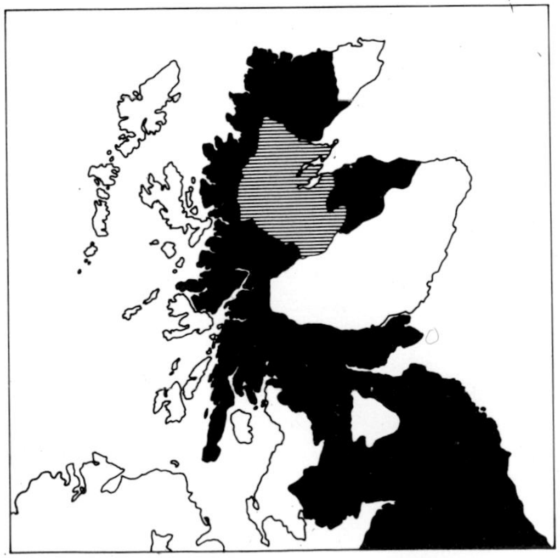

Distribution of Smooth-Newts in Scotland
Black areas=recorded *Stipple area=doubtful* *White areas=unrecorded*

Bibliography

BREADALBANE, MARCHIONESS OF (1935) *The High Tops of Blackmount*, Edinburgh.
CAMERON, A. G. (1923) *The Wild Red Deer of Scotland*, London.
DARLING, F. F. (1939) *A Naturalist on Rona*, Oxford.
DARLING, F. F. (1948) *Island Years*, London.
DARLING, F. F. (1937) *A Herd of Red Deer*, Oxford.
DARLING, F. F. and BOYD, J. M. (1968) *The Highlands and Islands*, London.
EDLIN, H. L. (1956) *Trees, Woods and Man*, London.
GODFREY, G. and CROWCROFT, P. (1960) *The Life of the Mole*, London.
HARVIE-BROWN, J. A. *et al.* (1887–1904) *Vertebrate Fauna Series*, Edinburgh.
MATTHEWS, L. H. (1952 and 1960) *British Mammals*, London.
MAXWELL, G. (1967) *Seals of the World*, London.
MELLANBY, K. (1971) *The Mole*, London.
NAHLIK, A. J. DE (1959) *Wild Deer*, London.
PEARSALL, W. E. (1950) *Mountains and Moorlands*, London.
RITCHIE, J. (1920) *The Influence of Man on Animal Life in Scotland*, Cambridge.
SCROPE, W. (1894) *Days of Deerstalking*, London.
SELBY, P. J. (1836) On the Quadrupeds and Birds inhabiting the County of Sutherland, *Edin. New. Phil. Jour.* for 1836: 156–61, 286–95.
SHORTEN, M. (1954) *Squirrels*, London.
SMITH, M. (1951) *The British Amphibians and Reptiles*, London.
SPEEDY, T. (1920) *The Natural History of Sport in Scotland, with rod and gun*, Edinburgh and London.
ST. JOHN, C. (1919) *Wild Sports and Natural History of the Highlands*, London.
ST. JOHN, C. (1884) *A Tour in Sutherlandshire*, Edinburgh.
STEPHEN, D. (1963) *Watching Wildlife*, Glasgow and London.
STEPHEN, D. and LOCKIE, J. D. (1969) *Nature's Way*, London.
THOMPSON, H. and WORDEN, A. (1956) *The Rabbit*, London.
WHITEHEAD, G. K. (1964) *The Deer of Great Britain and Ireland*, London.
WHITEHEAD, G. K. (1972) *Wild Goats of Great Britain and Ireland*, Newton Abbot.
MCVEAN, D. N. and LOCKIE, J. D. (1969) *Ecology and Land Use in Upland Scotland*, Edinburgh.
VAN DEN BRINK, F. H. (1967) *A Field Guide to the Mammals of Britain and Europe*, London.
VENABLES, L. S. V. and U. (1955) *Birds and Mammals of Shetland*, Edinburgh and London.

Acknowledgements

We gratefully acknowledge the use of material by the photographers listed below:

S C Bisserot pp 29, 49, 53, 54, 82, 95; Jane Burton pp 31, 42, 43, 81;
Ake Wallentin Engman p. 18; Russ Kline p. 52; Ernest G Neal p. 63;
N Picozzi p. 29; Hans Reinhard pp 17, 38; A Tewnion pp 37, 38.

Photographs and map of newts p. 109 are reproduced from the book "The British Amphibians and Reptiles", Collins New Naturalist Series, by Malcolm Smith; photographs by W S Pitt.

The map showing the distribution of wild goats in Scotland is reproduced from the book "The Wild Goats of Great Britain and Ireland" by G Kenneth Whitehead, published by David & Charles.

All photographic material in the book, other than that listed, is the work of the author David Stephen.